IN SEARCH OF A FAIR WIND
The Sea Letters of Georgia Townsend Yates, 1891-1892

By Georgia Townsend Yates and Clint Cargile
Zea Mays Publishing | Sycamore, Illinois

IN SEARCH OF A FAIR WIND

The Sea Letters of Georgia Townsend Yates, 1891-1892

Copyright © 2017 by Zea Mays Publishing. All rights reserved.

No part of this book may be reproduced in any manner whatsoever without written permission except in the case of brief quotations embodied in articles and reviews.

For information, address Zea Mays Publishing, P.O. Box 218, 308 West State Street, Sycamore, IL 60178.

Edited by Kelly Unger
Design/layout/imagework by H. Neu, SimpsonPointPress.com
Cover and map designs by Andrew Higdon

ISBN-13: 978-1545230503
ISBN-10: 1545230501

ZEA MAYS
PUBLISHING

Oh! for a fair wind!! I shall never again hope to be anywhere in any given time; when I am *there* I will begin to think about getting there – not before.

—Georgia Townsend Yates, on board the *Willie Reed*
February 28, 1892

Table of Contents

Introduction — 1

Part One: Jack and Georgia — 5

Part Two: The Letters — 55

   Chapter 1 — 65
   Sailing to Japan: September 8, 1891–March 15, 1892

   Chapter 2 — 106
   Japan: March 19, 1892–April 9, 1892

   Chapter 3 — 117
   Singapore: April 15, 1892–August 6, 1892

   Chapter 4 — 143
   Return Home: August 15, 1892–December 9, 1892

Part Three: Life After the Letters — 161

Appendices

   Appendix A — 202
   Family Trees

   Appendix B — 205
   Sea Shanties

   Appendix C — 208
   A Tale of Two Transcripts

   Appendix D — 211
   1912 Political Speech by Captain John E. Yates

   Appendix E — 215
   A Trip to Round Pond, Maine: From Amos Claycomb's 1902 Diary

Bibliography — 222

Acknowledgments — 227

About the Author — 229

Introduction

The story of Georgia Townsend Yates comes to us primarily via a collection of letters that she wrote to her mother from 1891 to 1892 while on board her husband's ship, the *Willie Reed*. She and her husband, Captain John Elvin Yates, had embarked on a fifteen-month journey to Japan and Singapore. With them were their infant daughter, Dorothy, and a crew of twenty-eight men. This was Georgia's first and only sea expedition. It was also the last sea expedition for Jack.[1]

Georgia Townsend was the scion of the Townsends and the Pierces, two prominent pioneer families of Sycamore, Illinois, who had sought the fair wind that carried settlers from the East to the western frontier. The Townsends made a name for themselves in agriculture and in their honorable support of the Underground Railroad. The Pierces had earned a fortune in real estate investment and finance. Georgia's family helped shape the political and economic future of her hometown. To this day, the name of Daniel Pierce, Georgia's grandfather and founder of one of Sycamore's most successful banks, remains strongly associated with the business and financial history of DeKalb County, Illinois.

Because of her family's significant role in building and shaping Sycamore, much of Georgia's family history had been researched and recorded long before this project began. Jack Yates, however, was a different story. His background and history remained largely unknown outside of Boise, Idaho, where late in life he played a significant role in building that city's downtown and business community. While in Boise, he kept the honorific "Captain" but also earned several other titles: politician, banker, capitalist, philanthropist, orchardist, and "the sheep man."[2] Jack Yates, it appears, was a man of diversified skills and interests who could adapt to any environment. He didn't move his family to Boise until he was in his

---

[1] Captain John Elvin Yates was known to family and friends as Jack or Captain Jack. Georgia referred to him as Jack throughout her letters, so I will refer to him as the same throughout this book.
[2] *Idaho Daily Statesman*, Mar. 4, 1902.

fifties, proving that at any age he could follow the wind of opportunity. Unfortunately, little is known of his life before then. I have tried to uncover that history by tracing his early life both in his hometown, Round Pond, Maine, and at sea.

Georgia's original letters no longer exist (or have not been located), but they were transcribed and then passed down through generations of relatives still residing in her hometown of Sycamore, Illinois. When I received these letters from Doug Roberts, great-grandson of Georgia's brother, Fred Townsend, I was instantly intrigued by Georgia and Jack's story. She was the daughter of an affluent midwestern family. He was a dashing, successful East Coast sea captain twenty-one years her senior. She had never been married. He had recently been widowed. How and when did they come to meet? Why did they marry? What compelled Georgia to leave her privileged world behind to join her husband for fifteen months at sea (and with an infant, no less)?

But what truly intrigued me about Jack and Georgia's story were the letters themselves. They offer a glimpse of a common but little-known aspect of maritime history: captains' wives accompanying their husbands on long sea voyages. The letters recorded Georgia's observations not just of life at sea, but also of the trials of raising an infant aboard a close-quartered, ocean-locked vessel—a world unto itself.

Fortunately for the modern reader, Georgia's writing was crisp, witty, and entertaining. At times she made wonderfully succinct observations, describing waves "hop-skipping over the deck" or the ship sailing through "a sea of melted silver." Other times, she showed that she could be quite clever and funny. Some examples:

> **On life at sea:** "I do not yet understand the art of balancing myself and I came near standing on my head in the tub two or three times."
>
> **On Dorothy:** "She climbs and wriggles all the time and is withal so sweet that it's all I can do to keep from eating her."
>
> **On crossing over the equator:** "Dorothy and I slept right through it and never felt the bump."
>
> **On the ship's steward:** "He won't work, and as he is not ornamental he is about as useless as he could be."

THE SEA LETTERS OF GEORGIA TOWNSEND YATES, 1891-1892

I have divided this book into three parts. Part One provides an extensive biographical sketch of both Georgia and Jack before the voyage. While I believe the letters stand alone and can be enjoyed without this background material (which may appeal most to local historians and genealogists), I want the reader to learn who these people were before they embarked together on a journey halfway around the world. This section begins with Georgia, on whom there was considerably more information, and provides background on her ancestors and the towns where they lived. I have put extra emphasis on Georgia's mother, Eleanor Pierce Townsend, because she was the single greatest influence in Georgia's life and also because Georgia's sea letters were addressed to her. To understand these letters, it is important to understand the relationship between Eleanor and Georgia. Part One then moves to Jack and his family in Maine, and it concludes with Jack and Georgia's brief time together before setting off on their journey.

Part Two is comprised of the letters themselves, fully footnoted to provide additional information and backstory, define terminology, and clarify Georgia's casual references to people and places familiar to her and her mother. Some of the footnotes repeat information from Part One for those who choose to read only the letters. For clarity and ease of reading, I have included maps of their journey and divided the letters into four chapters: chapter 1 covers the trip to Japan, chapter 2 covers the time Jack and Georgia were in port in Japan, chapter 3 covers the trip to and time in Singapore, and chapter 4 covers the voyage home. I interrupt chapter 1 with newspaper accounts of a brief mutiny that took place on the *Willie Reed* during Georgia's voyage. I include these reports because Georgia provided sparse details about an incident that would have been the most thrilling and terrifying of her voyage and was widely reported on in American newspapers.

Part Three looks at Jack and Georgia's lives after the voyage. It surveys their time in Illinois and Maine and then their new life together in the West. End matter includes basic family trees, lyrics to sea shanties Georgia heard on her voyage, a 1912 political speech delivered by Jack when he was running for state treasurer of Idaho, and an excerpt from the journal of Georgia's nephew, Amos Claycomb, describing a trip he made to Jack's hometown of Round Pond, Maine, in 1902.

As much as possible, I have tried to use primary sources concerning

Jack and Georgia and their immediate family. Most of the information concerning their day-to-day activities came from newspapers, specifically the *True Republican* from Georgia's hometown of Sycamore, Illinois, and the *Daily Statesman* from Jack and Georgia's later hometown of Boise, Idaho. For distant relatives or those not crucial to the overall narrative, I often turned to secondary sources such as family members' compiled genealogies or family trees found on various genealogy websites. I believe genealogists will find this book to be a good reference and a starting point for further research.

Overall, this book is but a brief survey of two eventful lives. My main hope is that Georgia's letters will prove useful to historians and entertain and fascinate casual readers, all while honoring the memory of Georgia Townsend Yates, a spirited small-town girl who braved storms, mutiny, tedium, and tides to traverse the world in search of her own fair wind.

—*Clint Cargile*

Part One:
Jack and Georgia

Georgia Townsend and John Elvin Yates in 1889. (Joiner History Room)

An Interesting Social Event

On Thursday evening, November 7, 1889, twenty-three-year-old Georgia Wild Townsend of Sycamore, Illinois, third daughter of an affluent local banking family, married forty-four-year-old Captain John Elvin Yates, a recently widowed sea captain from Round Pond, Maine. They had known each other about three months. A brief account in the local newspaper, the *True Republican*, described the occasion simply as "an interesting social event."

The nuptials took place at the Townsend family home, a stately mansion that loomed over the neighboring houses on Sycamore's prestigious Somonauk Street. The ceremony itself was small, a private affair attended by relatives only, which was uncommon for a family of the Townsends' stature. When Georgia's sister Anna married six years earlier, the *True Republican's* lengthy announcement described the wedding as "one of the most charming social events" of the season, attended by nearly two hundred guests, including family and friends who had traveled from several neighboring communities and states. As was the custom of the time, the article described the decorations and provided a long list of the presents given to the bride and groom. Similarly, the wedding announcement of Georgia's cousin, Daniel Pierce Wild, described how the event brought out "the splendid assemblage of the wealth and fashion of the city. . . . The pavements rattled with a merry din as carriages and wagonettes hurried their passengers to the reception."

The short and simple announcement for Georgia's wedding lacked both the details and the flowery language of previous family nuptials. One must question, then, if there was an ambiguity behind the words "interesting social event."

After the wedding, the couple took a brief honeymoon—location unknown—and then returned to Sycamore in time for Thanksgiving. They stayed at the Townsend home through the spring (Georgia's mother was wintering in California, having left immediately after the ceremony). In May 1890, they traveled to Captain Yates's hometown of Round Pond, Maine, where they would make their permanent home. They returned to Sycamore the following winter, where Georgia gave birth to their first child, Dorothy, on January 30, 1891.

Always answering the call of his restless spirit, Captain Yates stuck around Sycamore only six weeks after his daughter's birth before embarking on a two-month journey into the West. He was looking to invest in some frontier land. He took along with him Georgia's uncle, William H. Townsend, and left his wife and newborn daughter in the care of Georgia's mother. Both men purchased large tracts of land just outside Boise, Idaho. The trip took Jack all the way to Washington and down the Pacific Coast. Soon after his return, he took his wife and infant daughter back to Maine to prepare for a sea voyage halfway around the world. Georgia and Dorothy accompanied him on this voyage, at which time Georgia wrote the letters that are the foundation of this book.

But these details are simple names and dates, collected mostly from local newspapers and a few scattered family documents. They fill in some blanks, but generate more questions than answers. Who was Georgia Wild Townsend? Who was Captain John Elvin Yates? How did these two people from disparate regions, generations, and backgrounds come to meet, marry, and sail the world together? While Part Two of this book focuses on Georgia's letters, it is important to know who these people were before the voyage and which people, places, and events shaped them.

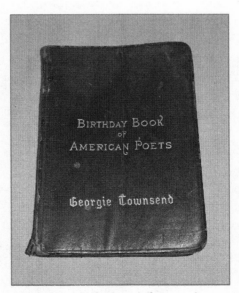

Georgia was known affectionately to her family as Georgie. Her nickname is embossed on the cover of her copy of the *Birthday Book of American Poets*. (Wendy Jones Smoke Collection)

Georgia Wild Townsend

Georgia Wild Townsend was born on Thursday, October 25, 1866, at her parents' farmhouse just north of Malta, Illinois. She was the fourth child and third daughter of Amos and Eleanor Pierce Townsend, the former hailing from prosperous farming stock, the latter the daughter of wealthy Sycamore financier Daniel Pierce. In the Townsend family, names often derived from close relatives, so Georgia was named for her uncle, George Parnell Wild, a successful Sycamore merchant who had married Eleanor's sister, Sarah Ann Pierce, in 1862.

At the time of Georgia's birth, Malta was a village "on the high tide of prosperity" that would "support a town of three or four times its present population."[1] But in 1872, a large fire swept through the village, destroying most of its businesses and greatly diminishing its growth and progress. Amos, who shared his father and father-in-law's keen instincts for land speculation, must have known that his future lay elsewhere. In 1876, Georgia's grandmother, Phebe Brundage Pierce, passed away, leaving sixty-two-year-old Daniel Pierce all alone on his large estate just outside Sycamore. Amos and Eleanor relocated there, and Amos took over the property's

The 1857 wedding portrait of Georgia's parents, Amos Townsend and Eleanor Pierce Townsend. (Joiner History Room)

---

[1] Henry L. Boies, *History of DeKalb County, Illinois* (Chicago: O. P. Bassett, 1868), 492–493.

management. As a result, ten-year-old Georgia spent her most formative years in the fast-growing community of Sycamore, a midwestern agricultural town quickly moving toward industrialization and gentrification.

Although details of young Georgia's life are sparse, her letters show us that she was clever, witty, well read, highly educated, extroverted, strong willed, and independent. She had a fondness for theater and liked to be the center of attention. She liked cats, one can assume, because she always had them. Georgia was, however, a self-confessed layabout and self-conscious of her appearance. In her letters she makes note of her weight ("quite plump") and her teeth ("large and far apart"), and she worries that her daughter "is going to look too much like me to ever be a beauty."[2]

While Georgia enjoyed a privileged upbringing, she and her siblings also lived in the public eye due to the family's high standing in a small, tight-knit community. Georgia grew up in a rigid stratum of society during the Victorian Era, so a certain decorum was expected of her; as a woman and daughter, she had little leeway to pursue her own interests. On top of all that, Georgia's family was a regular topic of society talk; they could barely step foot outside of their home—or leave town—without some mention in the local newspapers. However, Georgia inherited her mother's strong will. Eleanor Townsend made sure that her children were socialized, cultured, and educated. She raised her daughters in the Universalist Church, which carried a progressive view of a woman's role in society.

But to really understand the story of Georgia Wild Townsend and the family that shaped her, we must look back further, across the country to another small farming community nearly nine hundred miles away from Sycamore.

Neversink, New York

The "Neversink country," as it was known, is located in the rolling foothills of the Catskill Mountains, a hundred miles northwest of New York City. After the first white settlers trickled in during the 1740s, they were either driven out by Native Americans or they fled from recurring hostilities with the French and later the British. The first permanent white settlers arrived around 1788 and found a region brimming with natural resources. In one of those great ironies of history, the original village of Neversink (often referred to as "Old Neversink") now lies submerged beneath the Neversink Reservoir, victim of a 1950s government project to

---

[2] Georgia Townsend Yates to Eleanor Townsend, Nov. 7, 1891, and Nov. 22, 1891.

Georgia Wild Townsend was named after her uncle, George Wild, who was married to Eleanor's only sibling, Sarah Ann Pierce. (Joiner History Room)

supply fresh water to New York City. But at the turn of the nineteenth century, Neversink was home to a vibrant community of farmers, lumbermen, and cattle ranchers.

Georgia Townsend's roots ran deep in Neversink. All of her grandparents met, married, and thrived there. All but one of them were born there. Georgia's parents were born there, also, as were most of her aunts and uncles. The small party of pioneers that arrived in 1788 included one Tobias Hornbeck, Georgia's great-great-great-great-grandfather. Even the region's topography evokes her family. The two tallest elevations in Neversink are Denman Hill and Thunder Hill. Denman Hill earned its name from the Denman family of Sussex, England, early settlers in the region. Their line eventually led to Ann Denman, Georgia's paternal grandmother. Thunder Hill became the home of the Porter and Cargill families. Francis Porter, who settled his family in Sullivan County, was Georgia's great-great-grandfather. His neighbor, Abraham Cargill, was another of Georgia's great-great-grandfathers.

Georgia's maternal grandfather, Daniel Pierce, was born in Neversink in 1814. He was the fourth of six children and the youngest son. His father died in 1819 at the age of thirty-one, when Daniel was only five years old. His death left the family in distressed circumstances, forcing Daniel to take on work at a young age and miss out on a formal education. By the age of twelve, he was hiring himself out to farms across Neversink, a burden he endured until he had saved enough money to lease his own land.

In 1835, twenty-one-year-old Daniel married Phebe Jane Brundage, whose family had recently moved to Neversink from nearby Orange County. Daniel and Phebe had two daughters, Eleanor Jane (Georgia's mother) and Sarah Ann. As Daniel's family grew, so did his ambition. He began building up farms and leasing them to immigrants who could afford a move-in-ready operation. This was Daniel's first taste of land investment, and he proved quite adept at it. This skill would later serve him well in the West.

On the Townsend side of the family, Georgia's great-grandfather, Joshua Townsend, brought his family to Neversink in the early 1800s. There he had two sons, Stephen (Georgia's grandfather) and Charles. With the help of his two boys, he operated a small but successful farm.[3] But in 1836, when Joshua was nearly fifty years old, he turned his sights west and decided to pursue a new life on the nation's open frontier.

There are several reasons why men like Joshua Townsend chose to uproot their families and move west. Some were economic—better opportunities and cheap land—while some were personal—escape, a fresh start. Joshua was the fourth born of eight children, many of whom already had children and grandchildren. Neversink bulged at the seams with Townsends. With more settlers arriving daily, land grew scarce and many settlers felt overcrowded. The vast open prairieland of

Georgia's maternal grandparents, Phoebe Brundage Pierce and Daniel Pierce. (Wendy Jones Smoke Collection)

---

[3] It is possible that Joshua Townsend employed a young Daniel Pierce to help out on his farm during this time.

the Midwest beckoned.

But there was another force that pulled New Yorkers west: the opportunity to fight on the front lines against slavery. A small Quaker community operated in Neversink, and even many of those settlers who did not follow that faith still incorporated its quiet and peaceful creed and its aversion to slavery into their everyday lives. The Townsends were of this ilk. Slavery disgusted them, but there was little they could do to combat it while tucked away in the hills of southern New York. Illinois, however, shared a border with the highly contested slave state of Missouri and was a major route on the Underground Railroad. According to family lore, the Townsends looked west so they could swell the ranks of the abolitionists and help smuggle fugitive slaves to freedom in Canada. We will temporarily leave Daniel Pierce behind in Neversink as the Townsends move to a new land with new opportunities.

Georgia's paternal grandfather, Stephen Townsend. (Joiner History Room)

DeKalb County, Illinois

DeKalb County is a near-perfect rectangle—eighteen miles wide and thirty-six miles long—located in the heart of northern Illinois. It was named for Baron Johann de Kalb (1721–1780), a German-born French officer who lost his life fighting for the colonies in the American Revolution. The county seat, Sycamore, lies sixty miles directly west of Chicago. When the first white settlers arrived in the 1830s, they discovered a vast, unending prairie, occasionally broken by groves of oak, poplar, hickory, and maple, all of it irrigated by a network of creeks and non-navigable rivers. These features provided incoming settlers with fertile farmland, abundant fuel, a boundless supply of building material, and plenty of water. The county's largest water source, the Kishwaukee River, snaked its way through the future sites of Genoa, Sycamore, Mayfield, and DeKalb. The first settlers found the land along the river's banks "as fertile and beautiful a region as the sun shines upon."[4] Without a navigable river, however, settlers had to depend on rudimentary, mud-rutted roads to ship or receive their goods and supplies.

The first settlers churned the rich black prairie sod and received a modest harvest for their labors. A harsh winter and rain-soaked spring drove many back east. Those who stayed managed to scratch out a meager but sustainable existence.

In 1836, several railroad companies raced to connect Chicago to the Mississippi River, kicking off a land grab in northern Illinois. These companies promised great wealth to towns along their proposed routes. Settlers and investors snapped up the land, sending values soaring and bringing a new wave of enterprising pioneers and land speculators into the region. Many of the railroad companies, however, couldn't make good on their promises and either went bankrupt or simply disappeared. As a result, the land bubble burst, real estate values plummeted, and several investors lost everything. But this dark cloud had a silver lining. Those who hadn't lost their investments (or had never invested in the first place) could purchase the land for cheap.

This was the situation that drew Joshua Townsend to DeKalb County. He rode the distance from New York to Illinois on horseback, and then laid claim to forty acres of land in Mayfield Township, an undeveloped stretch of prairie just west of Sycamore. He then returned to his family in New York to prepare for the move west.

---

[4] Boies, *History of DeKalb County*, 37.

The following year, Joshua's son Charles and his family climbed into a covered wagon and made the six-week journey to their new home. They settled on a portion of the claim and began converting the empty prairie to farmland. Around the time of Charles's arrival, the Illinois General Assembly designated Sycamore as the seat of DeKalb County, even though the town had only one permanent building, a combination tavern, boarding house, and dry goods store. Most of the population lived in this structure.

At this time, DeKalb County was still the wild western frontier. In an area populated by wolves and prairie rattlesnakes, settlers lived rough, hardscrabble lives. They had to be on constant alert to defend themselves against bandits, horse thieves, squatters, and claim jumpers. Captured criminals still faced so-called frontier justice. But the real threat came from nature. Charles Townsend's family arrived at a time when "ague and bilious fevers were the prevailing diseases."[5] Floods were common, which encouraged mosquitos and more disease. Livestock spent the spring mired in mud only to freeze to death in the winter. Many settlers quickly gave up and returned home. Yet despite the harsh conditions, new settlers continued to pour into the region.

In 1840, Joshua Townsend returned to DeKalb County, carrying

Stephen Townsend's home in Mayfield, Illinois, just west of Sycamore. This depiction appeared in the 1871 *Combination Atlas Map of DeKalb County, Illinois.* The house stood until February 2017. (Sycamore History Museum)

---

[5] Boies, *History of DeKalb County*, 380.

three generations in a covered-wagon caravan. He and his wife were joined by their eldest son, Stephen (Georgia's grandfather), who brought along his wife and four children, including eight-year-old Amos (Georgia's father). When this second wave of Townsends arrived, the nearby village of Sycamore had grown considerably. It now boasted a population of around seventy people, mostly men. It had twelve homes, two stores, two hotels, a mill, a blacksmith shop, access to a stagecoach line, and one well. It had nearly completed its courthouse, which was also used as a school, multidenominational church, and public meeting house.

The Townsends all lived and worked together on the same farm, where Stephen and his wife, Ann Denman, had four more children (they had ten total; six lived to adulthood). Charles Townsend and his wife, Phebe Porter, also had ten children (eight lived to adulthood). The Townsend clan continued to expand the farm until it grew into several hundred acres. After many years, Joshua divided the land between his two sons. By the mid-1870s, the brothers owned over four hundred acres.

Underground Railroad

Henry L. Boies, a newspaper editor and DeKalb County's first historian, gave the following description of the county's role in the Underground Railroad:

> From the first settlement of the County, it had been the home of a strong, active, zealous party of anti-slavery men; men who were avowed abolitionists, who gloried in that name when it was a term of reproach; who not only voted for, but labored and expended their money for the freedom of the slave. Scattered here and there over the whole county, were numerous well-known stations on the "under-ground railroad;" homes of thrifty, hard-working, God-fearing haters of oppression, in which, it was well understood, the panting fugitive escaping from Southern Slavery, would be sure of finding rest, refreshment, a safe shelter, a warm welcome, and means to help him on to other stations on the route to what was then his only safe-guard, the flag of England on Canadian soil.[6]

If the opportunity to fight slavery and assist fugitive slaves did indeed draw the Townsends to the Midwest, then they remained true to their convictions. In DeKalb County, the family became outspoken abolitionists from the moment they arrived. Stephen Townsend and

---

[6] Boies, *History of DeKalb County,* 109. For a thorough examination of DeKalb County's role in the Underground Railroad, see Nancy Beasley, *The Underground Railroad in DeKalb County, Illinois* (Jefferson, NC: McFarland & Co., 2013).

his wife cofounded the Wesleyan Methodist Society of Mayfield, a denomination that broke ranks with the Methodist Episcopal Church after the latter refused to take a stronger stance against slavery. Stephen also identified with the Liberty Party, the short-lived anti-slavery party that formed in New York in 1840. When elements of that party evolved into the Freesoil Party in 1848, Stephen organized a local chapter. When the Freesoil Party merged with the Republican Party in 1854, he became an outspoken member of that party as well. He was "a very active and public-spirited man," noted DeKalb County historian Lewis M. Gross. "Very few men have exerted a larger influence over the community in which [they] lived than Stephen Townsend."[7]

But the Townsends did more than just speak out against slavery. Boies described the family as "neither ashamed nor afraid to be known as stationagents on the underground railroads." Boies also mentioned that there were "many an interesting story" to be told of their experiences. Unfortunately, he did not record those stories, nor have they been recorded elsewhere.[8]

In those days, the nearest market for Sycamore farmers was St. Charles (halfway between Sycamore and Chicago), so the Townsends operated a stretch of the Underground Railroad to that town. When delivering grain to St. Charles, they would hide slaves in a wagon with a concealed compartment at the bottom. After dropping off the slaves, the Townsends would return home while the slaves continued on toward Chicago.

The Townsends and all of the other families who took part in the Underground Railroad took an incredible risk. If caught, they could have faced fines, lawsuits, jail, or physical harm from bounty hunters or supporters of slavery. Fortunately, many of the settlers around Mayfield and Sycamore shared the Townsends' abolitionist views. At Stephen Townsend's funeral in 1883, Reverend George Young, Wesleyan minister, gave the following eulogy:

> Stephen Townsend, from his home on the prairie ever extended a helping hand to the black man fleeing from the chains of slavery to freedom from bondage, then only secure upon reaching Canada. Mr. Townsend advocated emancipation of slavery when ministers behind pulpits denounced it. He had courage to defy an unjust law and say to the slave catchers, that never by his help would he return a slave to the chains of a cruel master.[9]

---

[7] Lewis M. Gross, *Past and Present of DeKalb County, Illinois, vol. 1* (Chicago: The Pioneer Publishing Company, 1907), 155.
[8] Boies, *History of DeKalb County*, 59.
[9] *The DeKalb Chronicle Illustrated Souvenir* (DeKalb, IL: J. F. Glidden Publishing Company, 1899), 72.

In the early 1850s, Georgia's father, Amos Townsend, attended the Illinois Institute (later Wheaton College) in Wheaton, Illinois, about twenty-five miles west of Chicago. The school was founded by Wesleyan Methodists—like the Townsends—and was a known stop on the Underground Railroad. While there is no evidence that Amos took part in Underground Railroad activities while enrolled there, it is likely that he continued the family trade. It is also possible, due to the school's importance to the network between DeKalb County and Chicago, that his family placed Amos there strategically. Wheaton was, after all, directly between St. Charles and Chicago. Many slaves who passed through Wheaton likely had first passed through the Townsend farm outside Sycamore.

"Flush and Easy Times Everywhere"

While Amos was away at school, major changes swept across northern Illinois. During the early 1850s, the region experienced a massive agricultural boom. The promised railroads of the 1830s had finally appeared in Chicago, transforming it into a major transportation hub that provided the region's farmers with easy access to lucrative East Coast markets. An international affair also influenced the huge boom: the Crimean War in Europe. The conflict disrupted European supply chains and created a huge demand for American grain. The price of wheat jumped from $0.30 a bushel to $1.50. Farmers weren't just making a living; they were getting rich. With the profit from one acre of land, a farmer could buy ten more. Soon, the rustic log cabins that dotted the landscape gave way to large, cozy farmhouses.

These conditions created a new land rush. At the time, the Midwest's largely unsettled prairieland could still be purchased cheaply. The Townsends found themselves in a prime position to profit from this situation. Farmers and land speculators from the East poured into the region, as did the merchants who wanted to supply them. Many frontier towns experienced unprecedented growth. For the new settlers to prosper, however, they had to borrow from banks to purchase the land and then buy supplies on credit from local merchants. The merchants, in turn, had to borrow from the same banks to set up shop and offer goods on credit. No one seemed worried that DeKalb County residents held more debt than at any time in the county's existence. The price of grain continued to climb, and the market seemed secure. As one local historian put it, "it was flush and easy times everywhere."[10]

[10] Boies, *History of DeKalb County,* 419.

This economic situation attracted the attention of Daniel Pierce. In the two decades that the Townsends had struggled to build their fortune in Illinois, Pierce had become a prosperous businessman and land speculator back in Neversink. In the late 1840s, he had thrown his hat into the political arena and served three terms as Neversink's town supervisor. He'd also purchased several farms and invested in several businesses. Despite his success, this new land rush on the nation's western frontier presented a financial opportunity he could not pass up. After forty years in Neversink, Pierce sold off his investments, pulled up stakes, and pointed his wife and daughters toward DeKalb County, Illinois.[11]

In 1855, the Pierces settled in Mayfield Township, near the Townsends, where Pierce took up farming while buying and selling other farms and real estate.[12] At this time, Pierce's eldest daughter, sixteen-year-old Eleanor, met Amos Townsend, recent graduate of the Illinois Institute who had returned home to work on the family farm. It is not clear how the two met, but small frontier communities held regular social events and gatherings, and they would have crossed paths at any one of those. After only a year in Mayfield, however, Eleanor's father relocated the family again, this time to a large farm just southwest of Sycamore that would become the Pierce family estate. According to Boies, Sycamore was experiencing a time of "extraordinary prosperity and remarkable increase of population."[13] Pierce immediately set up an office in town as a money lender and real estate developer.

Eleanor and her sister, Sarah Anne, attended Sycamore schools and took an active role in the town's Universalist Church, forging a connection with that denomination that would last a lifetime. Sycamore's Universalist Church organized in the 1840s, landed its first permanent preacher in 1853, and built its first permanent church in 1854. When Eleanor and her sister arrived in Sycamore two years later, they were attracted to the church's progressive stances on slavery and the role of women in society.

But Eleanor was not long for Sycamore, either. On October 15, 1857, the eighteen-year-old married Amos Townsend and they settled on a farm southwest of Mayfield Township, two miles north of the village of

---

[11] It is unknown how much correspondence Pierce shared with former Neversinkers living in and around Sycamore, but so many had moved there that it is likely he would have used these connections to research his future home.

[12] It is unclear how well the Pierces knew the Townsends or whether the Townsends had any influence on Daniel Pierce moving his family to DeKalb County. Pierce was in his early twenties when the Townsends departed Neversink, so he would have been familiar with the family. Eleanor Townsend's obituary describes the two families as "old acquaintants" before coming to Sycamore. *Sycamore Tribune,* Dec. 23, 1904.

[13] Boies, *History of DeKalb County,* 417.

Malta, a budding community located on the Galena & Chicago Union Railroad. In Malta, Eleanor joined the local school board, becoming the first woman in DeKalb County to take such a position. She also cofounded that community's first Universalist Church Society. The couple remained in regular contact with both the Townsend and Pierce sides of the family, and much of their social life still centered on Sycamore. Eleanor made routine trips there to purchase supplies and visit family and friends. The families also gathered regularly for large dinners and to celebrate birthdays and holidays.

When the Crimean War ended, the grain-market bubble deflated, dropping demand for wheat, the region's most abundant crop. For most new settlers, their ability to pay their debt depended on the continued high price of grain. As a result, farms went under, businesses went bankrupt, and the banks followed. The entire country plunged into a great depression. To add to the distress, drought and blight descended on the region and destroyed many of the crops. *The DeKalb County Sentinel*

The farmhouse on Daniel Pierce's estate. This depiction appeared in the 1871 *Combination Atlas Map of DeKalb County, Illinois*. Due to the way the townships were divided, the Pierce property stood in Cortland Township, even though it was located just outside Sycamore and was closer to Sycamore than the village of Cortland. At the time of this writing, the site is occupied by a Blaine's Farm & Fleet. (Sycamore History Museum)

The Townsend family. Stephen sits in the center with his wife, Ann Denman Townsend, to his left. Amos and Eleanor sit to the right. This family portrait was taken shortly after Amos and Eleanor's 1857 marriage. (Wendy Jones Smoke Collection)

reported on the situation: "These are indeed, hard times. Our citizens are badly in debt; there is scarcely one-fourth of a crop to be sent to market to raise money… We beg leave to ask in all sincerity, what are people to do?" Many recent settlers fled back east or moved further west, some sneaking off in the dead of night to avoid their debt.

The Townsends certainly experienced hardship as the region plunged into another depression, but because they were well established before the boom and would have carried little debt, they weathered the storm relatively unscathed. Likewise, Daniel Pierce came through with his business and reputation intact. According to one source, Pierce "always knew when to buy and when to sell a piece of real estate, and the fluctuations in the money market were generally foreseen by him."[14] He was saved by his prescience, his prudence, and also his capital; unlike many of the incoming settlers, he had arrived in Sycamore with cash in hand and set up shop without debt. He no doubt lost money when land values plummeted, but not enough to destroy his fortune.

In 1858, Sycamore's business leaders came up with a plan to boost business and raise the town out of its economic slump. Instead of waiting for a major railroad company to build a new line through Sycamore, they would raise the funds to build their own railroad. It would run five miles

---

[14] *The Biographical Record of DeKalb County, Illinois* (Chicago: The S. J. Clarke Publishing Company, 1898), 40.

south to the town of Cortland, where it would connect with the Galena & Chicago Union Railroad and provide Sycamoreans with access to Chicago markets. Due to a substantial boost from the town's largest newspaper, the *True Republican*, the railroad's board of directors raised the necessary funds in only a few weeks and began construction by the end of the year.[15] Stephen Townsend sat on the first elected board of directors, as did Charles O. Boynton and Moses Dean, two New York transplants (Dean was a fellow Neversinker) who had close business ties to Daniel Pierce.[16] The railroad was a huge success and brought growth and financial stability back to Sycamore. It also arrived just in time to fend off the economic disruption brought on by the Civil War.

While Sycamore built its first railroad, Amos and Eleanor welcomed the first addition to their family. In 1858, Eleanor gave birth to Frederick Brundage Townsend, the first of five children. Next came Jennie and Anna, born in 1860 and 1864, respectively. Then, at long last, came the subject of this book, Georgia Wild Townsend, born in 1866.[17]

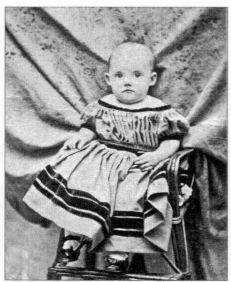

Georgia at six months. (Wendy Jones Smoke Collection)

When Georgia was born, Amos and Eleanor would not have been considered rich, but they were certainly well off, and Georgia's early life would have been more privileged than most in DeKalb County. Eleanor had at least two full-time domestic helpers, and Amos used a rotating mix of Swedish and Norwegian

---

[15] Some of the railroad directors owned the newspaper, and the railroad and newspaper offices were located in the same building, also owned by one of the railroad directors.

[16] Moses Dean was born in Neversink on January 27, 1815, six months after Daniel Pierce. He followed a similar path as Pierce, buying and selling various farms while investing in mills and a tannery. In 1856, he sold off his property and came to Sycamore, where he followed Pierce's business model: working as a farmer while amassing a fortune through loans and real estate investments. He eventually expanded his farming operation to 1,400 acres. As with the Townsends, it is unknown whether or how well the Pierces and Deans were acquainted before moving to Sycamore.

[17] The youngest child, Mary Corey Townsend, was born almost a decade later in 1875. She was given her middle name in honor of the Corey family, friends and neighbors of the Townsends.

immigrants to handle the farm's heavy workload.

Little is known about life on the Townsend farm, but we can assemble some details from Eleanor's 1867 diary (the only year from which such a diary has been located). Each entry begins with a short comment on the weather (e.g., "Cloudy and much warmer") followed by a brief summary of the day's events, usually no more than a sentence or two. Specific details are sparse. For example, one Saturday entry states only that a house servant "did Saturday work." Here is a sample entry for Monday, January 7, 1867:

> Very pleasant today, quite warm – did the house work today while Allie worked. The clothes dried very nicely. Amos went to DeKalb with a load of wheat.

The diary rarely mentions Georgia, who was less than a year old when it began. There are a few casual references to traveling with "the baby." Eleanor does mention her on April 4, 1867: "Georgia nearly sick today with cold and teething." She follows the next day with "Georgia quite

The four oldest Townsend children. From left to right: Georgia, Fred, Anna, and Jennie. (Joiner History Room)

sick," and on April 7 writes, "Georgia seems considerably better." From May 3 through May 10, Eleanor worked continuously to sew Georgia some short summer skirts. The only other references to Georgia also involve Eleanor making her clothes of some sort.

From the diary, we learn that the Townsends kept up an active social life. In this time of horse-drawn wagons and treacherously muddy and uneven roads, Eleanor nevertheless made daily visits to friends and relatives or received them herself. These visits still took place during the coldest, snowiest days of winter or the hottest, wettest days of summer. During her frequent trips to Sycamore, she would check out the latest fabrics brought in by the Sycamore & Cortland Railroad. She was fond of visiting photography studios to have photographs taken of her children.

The Townsends also took several trips out of town and were regular attendees at dances and parties. In February 1867, they hosted their own party, which drew fifty guests despite a bad winter storm that blocked the roads with heavy snowdrifts. The family frequently dropped in on the Coreys to the north, the Lloyds to the south, or Amos's sister, Kate Brundage, who lived in Malta and happened to be married to Eleanor's cousin, Fred Brundage.

To keep up with the burden of being a farmer's wife, Eleanor employed two house workers, Angeline "Angie" Allen and a woman named Alice ("Allie"). Alice focused mostly on housework, while Angie watched after the children, especially baby Georgia. Angie worked in the Townsend household for forty years and acted as nurse to all of Amos and Eleanor's children and several of their grandchildren.

Pierce, Dean & Company

After the Civil War (and after his granddaughter Georgia's birth), Daniel Pierce's investments continued to grow, and he began pursuing greater and more lucrative real estate transactions. But a post-war recession claimed many local banks and businesses, including Sycamore's largest bank, D. Hunt & Company.[18] If Sycamore wanted to keep pace with the times, Pierce believed, it needed a strong financial institution. So he stepped in and started his own. In 1867, he founded Pierce, Dean & Company with partner Moses Dean, another native of Neversink, and investor Richard L. Divine, who came to Sycamore in 1858 from Fallsburg, New York (about ten miles south of Neversink).[19]

---

[18] Later newspaper articles credited the loss of the bank more to the incompetence of the bank's managers than to the recession.
[19] Both Dean and Divine became mayors of Sycamore.

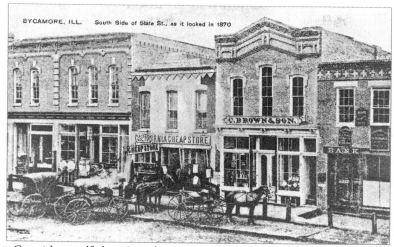

Georgia's grandfather, Daniel Pierce, cofounded the bank Pierce, Dean & Company in 1867 (seen on the right side of this postcard, ca. 1870). The bank created much of the Townsend family fortune. It existed under different names—most recently National Bank & Trust (NB&T)—until 2015, when First Midwest Bank purchased it. (Joiner History Room)

Just as the bank was getting underway, the country's economic conditions improved due to major investments in railroads and a second industrial revolution. While the nation shifted toward industry and manufacturing, Sycamore followed suit. Pierce, Dean & Company financed the shift just as western expansion renewed. Large factories cropped up along the Sycamore & Cortland railroad tracks. Within a few years, the bank had earned Pierce a substantial fortune, and he quickly became Sycamore's wealthiest citizen. *The Biographical Record of DeKalb County, Illinois,* described Pierce's bank as "one of the safest financial institutions in the state" and credited this to Pierce's "well-known conservative character and strict integrity."[20]

The Townsend Children

As stated earlier, in 1872 a large fire swept through Malta, Illinois, destroying most of its businesses and checking its growth and progress. The fire, combined with the explosive growth of nearby Sycamore and DeKalb, ensured that Malta would never grow into an economic powerhouse in the region. Despite this downturn, Amos and Eleanor stuck to their farm. In 1875, Eleanor gave birth to the youngest Townsend child, Mary Corey Townsend. That same year, the oldest child, Fred, departed for college at

---

[20] *Biographical Record*, 40.

Sycamore's Universalist Church moved to its new building on State Street in 1875, a year before the Townsends moved to Sycamore. Georgia's mother served as a trustee. Georgia took part in several children's and young people's programs, including dances and plays. When the church relocated in 1927, the city repurposed the old building as the Sycamore Community Center (pictured here, ca. 1930). Since 2004, it has been home to the Midwest Museum of Natural History. (Joiner History Room)

Lombard University in Galesburg, Illinois, where he studied to become a pharmacist.

When the Townsends joined Daniel Pierce on his estate in 1876, Amos took over its management and also joined his father-in-law's bank as a junior business partner. After twenty years of running his own successful agricultural operation, Amos quickly took to white-collar life and soon moved up to bank manager.[21] An early biography described him as "a thorough business man, methodical in all his ways."[22]

Eleanor was glad to return to Sycamore. Not only was it the social and cultural center of DeKalb County, but living there put her closer to her immediate family and in-laws. She also could return to her old Universalist Church, which had built a grand and imposing new building on State Street the previous year. Eleanor and her daughters immediately immersed themselves in the church. Georgia and her sisters took part in youth organizations while Eleanor served on several planning committees. Eleanor eventually served as director of the Ladies Aid Society and spent four years as president of the church parish organization (1887–1891). She

---

[21] When the family moved to Sycamore, their farm in Malta had grown to 651 acres valued at $32,550.
[22] *Biographical Record*, 163.

also represented the church at regional and national conventions, including events in Brooklyn, New York, in 1885, and Akron, Ohio, in 1886.

For the Townsend girls, the Universalist Church stood at the heart of Sycamore's social, civil, and religious life. It sought to enrich the community by organizing regular lectures, concerts, fairs, festivals, and socials. The church didn't limit its guest speakers to religious topics; orators lectured on the important political, scholarly, and humanitarian topics of the day. Events were designed for members of all ages, and Eleanor and her daughters played a key role in organizing them. The *Sycamore Tribune* summarized Eleanor's service to the Universalist Church, describing her as its "constant attendant and chief supporter. Always ready to perform any service in its behalf, willing to take up the work patiently and cheerfully… In short, she was a power in the church, devoted, resourceful, energetic, a natural leader, greatly beloved and guided at all times by the highest ideals."[23] The *True Republican* noted that "her church was almost as dear to her as life itself. No request in its behalf ever passed unheeded; but each was granted with encouraging words and a liberal hand."[24]

When Georgia and her sisters first arrived in Sycamore, the biggest change they faced would have been the transition from rural farm life to Sycamore high society. Thanks to their grandfather, they were the wealthiest family in town. Little is known about their day-to-day activities, but snippets can be gleaned from Sycamore's local newspapers; the family's elevated social status made their activities privy to the personal section of each issue. They were invited to a never-ending succession of social events: balls, dances, masquerades, and numerous church-related activities. One can assume that the sisters had an inexhaustible supply of male suitors.

The Townsend girls also had to transition to a life of leisure and travel. Freed from routine farm work, they took long vacations, often to visit old friends and relatives in New York or to summer on the East Coast. Amos and Fred joined them on these journeys, but for only a few days at a time, because Amos had to tend to his duties as bank manager and Fred had graduated from college and was now working at the bank as a bookkeeper and cashier. Without men around, family vacations became a strong bonding time for Eleanor and her daughters.

One by one, the girls went off to college, all attending Lombard University in Galesburg, Illinois. Eleanor had taken a special interest in the school,

---

[23] *Sycamore Tribune*, Dec. 23, 1904.
[24] *True Republican*, Dec. 24, 1904.

Eleanor with her daughters, ca. 1885. From left to right: Anna, Georgia, Eleanor, Mary, and Jennie. (Joiner History Room)

Lombard University in Galesburg, Illinois, ca. 1865. Eleanor was a major donor, and all of her children attended school there. Georgia was the only Townsend daughter who did not meet her husband at the university. (Library of Congress)

Sycamore's youth often took part in masquerades. Georgia helped organize several such events through the Young People's Society of Sycamore's Universalist Church. (Wendy Jones Smoke Collection)

which the Universalist Church had founded. She became a trustee and one of the school's most generous benefactors. She was known to be intensely involved in the school's affairs. Each year she attended the graduation ceremonies to present the Townsend Award for Excellence in College Oratory—commonly known as the Townsend Prize—which she both founded and funded. Two prizes were awarded annually to female students exhibiting exemplary oratorical skills. She also struck up a friendship with the school's president, Nehemiah White, who became a frequent guest at the Townsend home and sometimes preached at Sycamore's Universalist Church.

Jennie Townsend graduated from Lombard in 1880. Anna Townsend attended Lombard for one year (1879–1880) and later graduated from Rockford Female Seminary. Georgia attended Lombard in 1882 and graduated around 1886.[25] The youngest daughter, Mary Townsend, graduated from Lombard in 1894.

In 1881, Jennie married Charles A. Webster, a businessman from Galesburg whom she had met at Lombard University. The school's president presided over the lavish ceremony, which Eleanor had organized on the family estate outside Sycamore. After the honeymoon, the newlyweds made their home at Galesburg, where they became prominent members of that community's Universalist Church.

In 1882, several of Sycamore's young people threw a surprise party for sixteen-year-old Georgia at her family home. By this time, Georgia had become "widely and favorably known" in the community.[26] The following year, her sister Anna married Frank E. Claycomb, whom she also had met at Lombard. After the wedding, they too settled in Galesburg.

Over the next few years, the papers gave numerous accounts of Eleanor and her daughters visiting each other in their respective cities. During summers, teenage Georgia spent weeks at a time with her sisters in Galesburg. All the while, she remained active in Sycamore's Universalist Church, organizing social events and acting in plays. In the church's production of the comic opera *Il Jacobi*, she played "Sophronia Skeggs, a young lady of fortune and feeling."[27] Later, in her sea letters, she criticizes her abilities as a seamstress, but at the 1886 Sycamore Fair she won first prize for a handmade sofa pillow.

After Georgia's return from college, her father's health began to decline. Amos Townsend sought treatment at a sanitarium in Battle Creek, Michigan, making several trips there with Eleanor, but doctors could not determine what was wrong with him. With his wife and five children at his side, he died at home on August 25, 1887, at the

Georgia's father, Amos Townsend, passed away on August 25, 1887, when she was twenty years old. (Joiner History Room)

---

[25] The exact date of Georgia's graduation is unknown.
[26] *True Republican*, Nov. 9, 1889.
[27] *True Republican*, Jan, 26, 1887.

Georgia in her teens and early twenties. (Joiner History Room and Wendy Jones Smoke Collection)

age of fifty-four. Georgia was twenty years old.

With the passing of her husband, Eleanor decided to move her two daughters into Sycamore proper. In the summer of 1888, she bought a three-story mansion on Somonauk Street, a north-south corridor off Sycamore's downtown that housed the city's wealthiest residents.[28] The mansion stood on a two-acre lot at the intersection with Edward Street and was, at the time, the largest home in Sycamore.

Former mayor Richard L. Divine had built the mansion from 1877 to 1878. At the time, the *True Republican* called it "the finest private residence

---

[28] Many of the original houses on Somonauk Street still exist as part of the Sycamore Historic District. The Townsend home, however, was converted to a hospital and later removed (see pages 179-180).

The Townsend residence in Sycamore in 1889. The house was built from 1877 to 1878 by Sycamore banker and lawyer Richard L. Divine, a former business partner of Daniel Pierce. Eleanor purchased the house in 1888.

in this or any of the adjoining counties."[29] It would later describe the home as "a generally imposing style of architecture…ideally located in one of the choicest residence neighborhoods… No building in the city is more substantial."[30]

Immediately after purchasing the home, Eleanor organized a series of renovations to be completed while she took Georgia and Mary to the East Coast for the summer. Upon their return, the newly remodeled residence awaited.

While it is unknown how Georgia felt about this move, we can speculate that she, her mother, and her sister would have been thrilled to be living in the heart of Sycamore. They were a short walk from downtown, the train depot, a popular riding park, and, most importantly, their church. Their wealthy friends surrounded them, and they lived in a house built for guests and entertainment.

At the end of the year, Georgia's sister Anna Claycomb and her family moved from Galesburg to the Pierce farm outside of town. While Georgia was no doubt excited to have her older sister return to Sycamore, Anna's

---

[29] *True Republican*, Sept. 15, 1877.
[30] *True Republican*, Dec. 15, 1909.

move was precipitated by poor health, which she had suffered since the birth of her third child the previous year. Meanwhile, Daniel Pierce, now in his seventies, decided to step down as bank president and pass on the reins to someone younger and more capable. His first choice had been his son-in-law Amos, but after Amos's unexpected death, Pierce passed the torch to his grandson, Fred Townsend.[31]

Eleanor Townsend, who had devoted so much of her time in recent years to her family, found that her move to Somonauk Street reinvigorated her service to the community. She served on the local advisory board for the American Educational Aid Association and organized the Sycamore chapter of the Illinois Equal Suffrage Association. She also threw herself into charitable work, visiting Sycamore's poor, sick, and elderly and using her wealth to assist them. She had a knack for discovering "the lonely in the community" and opening her doors to them, including giving generous access to her private library. The local papers described her as "an angel of mercy."[32]

Georgia's brother, Fred Townsend, took over as president of their grandfather's bank. (Joiner History Room)

Eleanor was the central figure in Georgia's life. She loomed large in both the family and the community (especially her church). Like many members of her family, she was a smart, progressive thinker, who was very much ahead of her time. She fought for women's rights at a time when it was difficult to do so. She pushed for women's education—especially her daughters'—when people didn't see it as necessary. Georgia always admired her mother's efforts and looked up to her as the perfect role model. Eleanor, in turn, instilled in Georgia a confident, adventurous spirit. While the family wealth

---

[31] Some sources claim that Daniel Pierce had stepped down earlier and already named Amos Townsend his successor, with Fred Townsend taking over only after his father passed away in 1887.
[32] *True Republican*, Dec. 24, 1904.

enabled Georgia to travel and become educated, it was Eleanor who instilled in Georgia a desire to learn and to experience the world.

On July 24, 1889, the following announcement ran in the local news section of the *True Republican*:

> Mrs A. W. Townsend and daughter, Georgia, leave this week for the coast of Maine, where they will spend the summer.[33]

Two months later, they returned to Sycamore. Georgia's future husband soon followed. What happened while Eleanor and Georgia visited Maine is unknown and will be deliberated on later, but here we must take temporary leave of Sycamore and the Townsend clan. We must travel twelve hundred miles to a small seaside village tucked away on a deep harbor on Maine's middle coast: Round Pond, the home of Captain John E. Yates.

Eleanor Townsend around the time she and Georgia traveled together to Maine. (Joiner History Room)

---

[33] *True Republican*, July 24, 1889.

John Elvin Yates

Maine has over 2,500 miles of shoreline. The ocean's relentless waves and currents have carved the coastline into a series of peninsulas and points, harbors and islands. Long before Europeans settled the region—before the pilgrims arrived at Plymouth Rock—Maine's rich fishing resources attracted an undaunted ensemble of fishermen. These men built permanent fishing stations, which processed the fish (salted or pickled) and repaired and outfitted ships. These stations also facilitated a burgeoning fur trade, which grew inland. Demand for fish and fur brought in shipbuilders. Shipbuilders needed lumbermen to harvest the region's abundant oak, white pine, and spruce, which were ideal ship materials. These industries needed year-round workers. Workers needed to be fed, so farming took hold. Farmers brought their wives and children. Now young seamen, fur traders, shipbuilders, and lumbermen alike had access to women for marriage. This is how Maine was settled.

When Jack Yates was born in 1845, Maine's shoreline was stippled by small, isolated sea towns separated by rivers, estuaries, and swamps. If a town had a deep harbor, it was good for shipbuilding. If a town had a shallow harbor, it was good for fishing. Whatever a town's main industry, it always took pride in the number of sea captains it produced.

On Maine's middle coast, on the eastern shore of the Pemaquid Peninsula, lies a little cove called Round Pond. For generations it was the lifeblood of the eponymous village that surrounds it. Round Pond is one of several villages that comprise the town of Bristol, which is why historical records list Jack Yates's birthplace as either of these two locations. Round Pond was built on a deep and naturally protected harbor, which supported a strong shipbuilding industry in the late eighteenth and early nineteenth centuries. If men didn't find gainful employment in the shipyards, they worked as fisherman or signed on to a ship's crew. Round Pond produced several sea captains, and many of those sea captains bore the surname Yates.

The first of Jack Yates's ancestors to settle in Round Pond was his great-great-grandfather James Yeates (1700–1793), a sea captain whose father, John Yeates (1672–1724), had emigrated from Yorkshire, England, and settled in Rhode Island. James was born in

Rhode Island, where he married Jane McNair (1718–1802), a native of Ireland.[34] The couple eventually relocated to Round Pond. They had ten children: eight daughters and two sons. This generation appears to be the one that dropped the first "e" from Yeates. Our subject hails from the second of the two sons, Samuel Gates Yates (1755–1864), who married Margaret Johnston (1761–1809), the daughter of Scottish immigrants. They had ten children, including Jack's grandfather, George Yates (1789–1841), who also had ten children, one of whom was John Yates, Jack's father. Many Yates men, including Jack's father,[35] took to the sea.

John Yates joined the crews of many of the ships sailing in and out of Round Pond. He married a Round Pond native, Sophia Blunt, on June 2, 1842.[36] Their first child, Oscar Samuel Yates, was born July 15, 1842.[37] Our subject, John Elvin Yates, was born February 4, 1845.

Little is known of Jack's early years. Most of what comes to us appears in biographical essays written later in his life, after he had achieved success as a businessman in Boise, Idaho. One such essay described him as being "descended from a race of sailors" and noted that he "first saw the light of day within sight of the sea."[38]

Tragedy struck in 1849 when Jack's father was lost at sea, leaving Sophia a widow with two young sons. With the father's death, the family also lost its sole means of income, and Sophia and her two boys had to depend on the assistance of relatives. Fortunately, as we see from the family history and the abundant number of children born to each generation, an ample supply of relatives resided in the area. At the time of the 1860 census, Jack was living with his uncle, Robert Blunt.

---

[34] Also recorded as Jeanne and Jenny.

[35] The historical record contains conflicting information as to the name of Jack's father; he appears as both John and George. I have chosen to go with John, because most sources use that name, including a 1909 biographical essay of Jack Yates—written while he was still alive—and the 1909 death certificate of Jack's brother, Oscar. John is also the name used in the marriage listings of *The Genealogical Advertiser: A Quarterly Magazine of Family History, 1899, vol. 2* (Boston: Press of T. R. Marvin & Son, 1899), page 16, which lists a John Yates marrying Jack's mother, Sophia Blunt, on June 2, 1842, the service officiated by Samuel T. Hinds, Esq. However, Jack's obituary and death certificate list his father's name as George. It is possible that both names are correct; according to some versions of Jack's family tree, his grandfather's name is George John Yates. Perhaps Jack's father shared that name, which would explain the confusion.

[36] Some family genealogists have the marriage year listed as 1841. The source mentioned in the previous footnote recorded it as 1842.

[37] In historical records, Oscar Samuel Yates's name appears inconsistently. He is recorded as Oscar S. Yates, O. S. Yates, S. O. Yates, Samuel O. Yates, and Samuel Oscar Yates. Regardless, he went by Oscar and was known affectionately as Osc.

[38] *Sketches of the Inter-Mountain States: Utah, Idaho, Nevada, 1847–1909* (Salt Lake City: The Salt Lake Tribune, 1909), 313.

While little is known of Jack and Oscar's childhood, they would have spent considerable time fishing for porgies, exploring down on the seashore, collecting tube worms and sponges, or digging clams from muddy tidal streams. If they decided to wander inland, they could hunt for turtles, toads, snakes, and salamanders, all of which thrived in the moist Maine climate.

Jack received a common school education, but by the age of sixteen, he could no longer ignore the siren's call of the sea. His father and several relatives had lost their lives in answer to that call, but the sea was in his blood, as it was in Oscar's, who had already signed on to a ship's crew. Jack started out earning seven dollars a month. He steadily moved up in the ranks. In 1870, when he was twenty-five years old, he became the captain of the *A. L. Fitch,* a schooner named for US Civil War gunboat commander Amaya L. Fitch. The *A. L. Fitch* was built in 1867 in Bristol, Maine, and operated out of that same region, so it is possible that Jack had served as a mate on the ship before advancing to captain.[39]

Jack's flourishing career ran parallel to his brother's. Around the same time Jack took command of the *A. L. Fitch,* Oscar became captain of the schooner *Susan Stetson*. While serving as captains, Jack and Oscar met and married young women from prominent local families. Oscar married Delia O. Hoffses on March 31, 1872. Later that year, on August 28, Jack married Roxanna "Rockie" Ella Cox. The brothers then went into business together and commissioned two ships of their own, which they named in their wives' honor. The *Rockie E. Yates,* a three-masted schooner, launched from Damariscotta, Maine, in late 1872. The *Delia O. Yates* launched from the C. G. Merry shipyard in Damariscotta on October 11, 1873.

When the Yates brothers took to the high seas, they took their wives with them. At the beginning, these trips were nothing like the fifteen-month voyage Jack and Georgia would later take to Japan. The *Rockie E. Yates* stayed closer to home, plying up and down the Eastern Seaboard and into the Caribbean. The ship's most common routes took it from Portland, Maine, to either Boston, New York, or Philadelphia, and from there down to Cuba, Haiti, or the Bahamas. In good weather, the ship could make the trip from Havana to New York in seven days.[40] On most trips Rockie would

---

[39] It is unknown how long Jack captained this ship, though a newspaper maritime report lists a Captain Yates at the ship's helm in 1872. *Indianola* (TX) *Weekly Bulletin,* May 8, 1872.

[40] The *Delia O. Yates* followed similar routes as its counterpart, but it was a larger ship—registered for four hundred and fifty tons, as compared to the *Rockie E. Yates's* two hundred tons—so it added stops in Central America and parts of South America.

have been the only woman on board with a seven-man crew, but as soon as the ship reached a busy port, numerous American merchants and their families would have surrounded her. Jack and Rockie often spent weeks in port waiting for workers to reload the ship with cargo—more time than they spent at sea—so Rockie would have had plenty of opportunities to visit with other captains' wives. And she wasn't always the only woman on board. On some occasions another officer might bring his wife along, or

Postcard image of Round Pond, Maine, found among Georgia's family pictures. Date unknown. (Wendy Jones Smoke Collection)

Round Pond was known for its abundance of porgies. Georgia's brother, Fred, took this picture of one of its porgie processing plants in 1903. (Joiner History Room)

Rockie might invite a female friend or relative.

Life at sea was not always smooth sailing, however. The first year out on the *Rockie E. Yates,* while making a run from Damariscotta, Maine, to Hackensack, New Jersey, the ship lost its bowsprit "by running foul of" the bark *Scotland.* Jack had to put in at Boston for repairs.[41] The *Delia O. Yates* hadn't been on the water two months when several crew members got in a brawl with the crew of another ship in port at Boston. Two crewmen from that ship were beaten nearly to death, and several crewmen of the *Delia O. Yates* were arrested. Initial reports identified Oscar as being involved and subsequently arrested along with his men, but these reports turned out to be false. He wasn't on the ship at the time of the incident, and he was never considered a suspect. The false reports came about because one of the arrested crewmen also had the last name Yates. How or whether he was related to Jack and Oscar is unknown.

Over the next few years, the Yates brothers invested in more ships and their fortune continued to grow. Even through their successes, they still stuck to the sea as captains, sometimes hiring themselves out to ships other than their own. Newspaper maritime reports show that the Yates's own vessels were often under the command of other captains. For example, in official records, Jack is listed as the owner and captain of the *Rockie E. Yates,* but shipping arrivals and departures that ran in port-city newspapers show that a variety of captains commanded the ship over the years. It is difficult to determine which ships Jack might have captained at this time, because most maritime listings included only the captain's last

Captain John Elvin Yates. Date unknown. (Wendy Jones Smoke Collection)

---
[41] *Bangor Daily Whig & Courier,* Nov, 4, 1872.

name, and the state of Maine had disbursed countless Captain Yateses all over the world.

The years 1876 and 1877 delivered both fortune and hardship to the Yates brothers. On October 4, 1876, the C. G. Merry shipyard in Damariscotta, Maine, launched the *Hattie G. Dixon*, a new bark co-owned by the brothers. At 140 feet long and capable of carrying 630 tons, it was their largest ship yet. Jack took it out on its inaugural voyage, but Oscar was not there to celebrate with him. At the time, he had taken the *Delia O. Yates* on a rare trans-Atlantic voyage from New York to Liverpool, England.[42] He brought along his wife, who did not lack female companionship on this voyage; the ship's steward had brought his wife, also.

The ship arrived at Liverpool on October 15, discharged and loaded its cargo, and departed on November 2 for Cuba. The following account from the *New York Herald* tells what happened on November 4 while a native pilot steered the ship through the English Channel:

> The accident occurred at 7 o'clock at night. The schooner had all her lights burning and adhered strictly to the regulations governing the right of way of vessels on the high seas. The [British] steamer [*West Indies*] was discovered to keep her course regardless of the schooner, though hailed and signaled; and, when she struck the *Yates*, which sank in 10 minutes thereafter, afforded no assistance to the crew, though the wives of the captain and steward were on deck screaming for assistance. The pilot and captain of the schooner called several times to the officers of the steamer to throw a rope, in order to get the captain's wife and stewardess off, and no notice was taken of the appeal, and the pilot succeeded in climbing on board the steamer, where he seized a rope and threw it to the schooner, whereby the women were hoisted on board the steamer. The crew of the schooner then clambered up the sides of the steamer as best they could, and fortunately no lives were lost.[43]

The newspaper condemned the British steamer for its "exhibition of careless indifference to human life" and called for the officers to "meet with universal reprobation." It is unknown if any charges resulted from the incident. The loss of the *Delia O. Yates*, however, did not deter Oscar Yates or his wife from the sea. If anything, it encouraged the Yates brothers to build bigger.

Jack and Oscar's largest investment, the *Willie Reed*, launched at

---

[42] I have not uncovered evidence of a previous trip to England, so this may have been the only one, but there are several gaps in the newspaper records, so we cannot rule one out completely.
[43] *New York Herald*, Jan. 1, 1877.

Waldoboro, Maine, on June 11, 1877. Built by A. R. Reed & Company (and named for A. R. Reed's brother, William), the two-hundred-foot, three-masted schooner could carry 1,500 tons (more cargo than the *Rockie E. Yates, Delia O. Yates,* and *Hattie G. Dixon* combined).[44]

The ship was co-owned by the builder, investors in New York, and Jack and Oscar.[45] With a ship of this class, no port was beyond the brothers' reach. On shipping records, Oscar is listed as the ship's master, but he and Jack took turns captaining the *Willie Reed's* numerous voyages around the world, most likely to give each other respite from the long months at sea. While one brother was gone for a year or more, the other handled the shorter runs back home.

The *Willie Reed* arrived at an opportune time for the Yates brothers. In October 1877, just a few months after the ship launched, a "furious and destructive storm" struck Delaware Bay.[46] The *Rockie E. Yates,* anchored in the bay, was battered by the fierce winds and waves, beached, and severely damaged, putting it out of commission for several months. At the time, Jack was in command of the *Hattie G. Dixon,* so neither he nor Rockie were on board. Oscar was sailing the *Willie Reed* to Liverpool by way of the English Channel, undaunted by his loss of the *Delia O. Yates* the previous year. The damage done to the *Rockie E. Yates* was a setback to the Yates brothers, but the profits from the *Willie Reed* more than made up for the loss.

The *Willie Reed* must have presented a challenge to the Yates wives. While sailing between New England and the Caribbean (or even to South America), Rockie and Delia had learned to handle stretches of a week or two on the water. But accompanying their husbands on the *Willie Reed* meant spending over a year at sea, isolated with an all-male crew for months at a time. That could be the reason that sometime during the mid-1880s, Rockie asked a cousin to accompany her on one of these long voyages. The

---

[44] The launch of the *Willie Reed* was reported on June 12, 1877, in the *Portland* (ME) *Daily Press, the New York Herald,* and the *New York Tribune.*

[45] The *Willie Reed* was co-owned by the Yates & Porterfield Trading Company, an import-export business based in New York City. The company was founded in 1854 by Joseph W. Yates (1826–1904) and Charles Porterfield (1816–1899). Both Yates and Porterfield were born in Bristol, Maine, and took to the sea at early ages. They were both captains by their early twenties and remained so until they opened their trading company. In the company's early days, it did most of its business with West Africa, shipping goods to and from Liberia and Sierra Leone. It eventually expanded to include major ports around the world. Joseph Yates and Charles Porterfield retired in 1884 and 1885, respectively, so they had departed the company by the time Jack and Georgia sailed for Japan. Joseph Yates was distantly related to Jack and Oscar; he and their father were second cousins (they shared a great-grandfather: James Yeates). The family and hometown connection, along with the brothers' proven success, made for an enduring business relationship.

[46] *New York Tribune,* Nov. 6, 1877.

The *Willie Reed*, the ship that would carry Jack and Georgia Townsend Yates across the globe, launched from Waldoboro, Maine, on June 11, 1877. This photograph was taken at the launch. A description of the ship ran in the *Portland Daily Press* the following day: "She is built of white oak and hard pine; length of keel, 199 feet; depth, 26 feet; breadth, 40 feet. No pains or expense have been spared to make her first-class and one of the best vessels ever built here." (From The New York Public Library)

story has been passed down for years by the Sawyer family of Bristol, Maine, because it was on that fateful voyage that Rockie's cousin, twenty-four-year-old Mary Eliza Fossett, met and fell in love with the ship's first mate, Orlando Chester Sawyer. Soon after the ship returned to the United States, they married and started a family.[47]

---

[47] Mary Eliza Fossett was Rockie Yates's first cousin once removed. Mary's father, Thomas Fossett, and Rockie Yates were first cousins on her mother's side. As an interesting side note, after Mary Eliza Fossett passed away in 1895, Orlando Sawyer married a young woman named Rockie Yates Loud, who was also a first cousin of Rockie Yates. Rockie Yates Loud was born October 2, 1872, just over a month after Rockie and Jack married, and was no doubt named in honor of that special occasion. Also of interest, the oldest child of Mary and Orlando, Arthur Sawyer, went on to marry Verna Yates, daughter of Albert Yates, who

Unfortunately, little is known about the personal life of Jack Yates, his brother, or their wives. There are no known letters or journals. We can surmise that the brothers kept in regular contact with each other and their mother while away at sea, because Georgia's letters reveal that Jack wrote to them often and received letters from them at each port. We can also surmise that the Yateses had intermittent but active social lives during their voyages. From Georgia's letters, we learn that while in port, it was common for the captains and their wives to socialize with the officers of other ships or acquaintances on land. We also don't know how often the wives sailed with their husbands. Did they travel on every voyage or only once or twice a year as a vacation of sorts? Were there long stretches where they didn't accompany their husbands at all? All we know is what appeared in the newspapers, maritime records, and sparse family documents.

After fifteen years of marriage, Jack's first wife, Roxanna "Rockie" Yates passed away on January 1, 1888 (cause unknown). She is buried in Chamberlain Cemetery in Round Pond, Maine. (courtesy of Libby Harmon)

For example, we know that in 1884 Jack sold the *Rockie E. Yates* to Thomas Harris, who had previously served as the ship's captain. But we don't know why. Was he in financial trouble? Did Harris make an advantageous offer? Was the ship no longer worth Jack's time because it was the smallest one in which he held an interest? Regardless of the reason, the timing of the sale was fortunate for Jack, because one year later, while carrying a cargo of salt from Haiti to New York, the *Rockie E. Yates* sank off Ragged Island in the Bahamas.

Another unanswered question: Why did the Yates brothers not have children? When Jack married Georgia, they had a daughter fifteen months

---

was Jack and Oscar's cousin and worked as a crewman on some of their ships.

later. In all, they had eight children (one died in infancy). What prevented Jack and Rockie from having children?[48] Or Oscar and Delia? These are questions we will probably never answer.

We do know, however, that Jack lost his first wife on January 1, 1888. Roxanna "Rockie" Yates died one month after her thirty-seventh birthday, cause unknown. She and Jack had been married fifteen years. She is buried in Chamberlain Cemetery in Round Pond, Maine.

At the time of Rockie's death, the *Willie Reed* was in Australia under a Captain Yates, but it is unknown whether this was Jack or his brother. From there the ship went to the Philippines before returning to New York in May 1889 with a cargo of sugar and hemp. It was in New York for only a month before it set out again for a year-long voyage to the US west coast, Shanghai, Cuba, and back to Philadelphia. For this trip, however, we know that Oscar was at the helm, because Jack was still in Maine, about to meet his future wife.

---

[48] In the course of my research, I found a family tree compiled by one of Jack and Georgia's descendants that claimed that Jack and Rockie had two sons who were adults when Jack and Georgia met. I have never found evidence of these sons or mention of them outside this one document. Jack and Rockie were married only fifteen years, so adult sons would have been unlikely when Jack and Georgia met. The 1880 census shows that Jack and Rockie lived with Jack's mother, and there are no children listed. My conclusion is that a family genealogist mistook information about Jack Yates's father—they're both named John (see footnote 35 on page 36)—and recorded that the elder John's sons—Jack and Oscar—belonged to Jack.

Jack and Georgia

We don't know how they met. Perhaps it was at church (Jack and his mother were also Universalists). Perhaps it was a social function (his wealth would have put him in the same social circles as Georgia and Eleanor). Perhaps they shared an acquaintance (Round Pond was not a big community). All we know is that in late July 1889, Georgia and her mother traveled east to spend the summer on the Maine coast. Two months later, they returned to Sycamore and resumed their regular lives. Eleanor organized and hosted her usual social functions while Georgia took an active role in weekly devotional meetings for young people at the Universalist Church. One month after Eleanor and Georgia's return, Jack arrived in Sycamore as "the guest of Mrs. E. P. Townsend and family."[49] Two weeks later, on November 7, 1889, he and Georgia married in a private ceremony at the Townsend home.

The following wedding announcement ran in the *True Republican:*

> An interesting social event was the marriage, on Thursday evening, of Miss Georgia Townsend, daughter of Mrs. E. P. Townsend, to Captain John E. Yates, of Bristol, Maine. The ceremony was performed in the presence of relatives only at the beautiful residence on Somonauk Street. After an absence of a few days the bride and groom will return to Sycamore, where they will remain until Spring, when duty calls them to their future home in Bristol, on the coast of Maine, Captain Yates being engaged as owner and commander of a vessel. The bride is widely and favorably known in this vicinity, where she was born and brought up, and the best wishes of her friends will accompany them in their new relationship. They will be at home to their friends after November 26.[50]

According to the marriage certificate, they were united in marriage by Reverend John E. June, pastor of the Universalist Church. The witnesses were Mary Boynton, a long-time friend who would later marry Georgia's brother, and Mary C. Gerts, another family friend.

Interestingly, when the newspaper reported that Jack and Georgia had secured a marriage license (in the same issue as the wedding announcement), it gave Jack's age as thirty-eight, or fifteen years Georgia's senior. He was actually forty-four, or twenty-one years her senior. The newspaper could have made a simple mistake, or it could have been fed false information in an effort to conceal the couple's true age difference.

---

[49] *True Republican*, Oct. 23, 1889.
[50] *True Republican*, Nov. 9, 1889.

The next mention of Jack and Georgia in the local papers appeared on January 1, 1890, when the *True Republican* reported that they were spending the holidays in Galesburg with Georgia's sister, Jennie Webster.

Historical Conjecture

So let us pause in the chronology of these two lives and examine the questions that have cropped up concerning Jack and Georgia's story. This section is full of conjecture and speculation, but seeks to understand what brought this couple together.

To begin, why did only Eleanor and Georgia go on this trip to Maine? Fred did not go because, as was usually the case, he had to stay behind to work. At the time of the Maine trip he was thirty-one years old, president of a large and successful bank, and had been elected as a city alderman. The two older sisters, Jennie and Anna, did not go because they were married, had small children, and rarely took part in their mother's long excursions. In addition, Anna had just given birth to her fourth child and was in poor health.[51] This just leaves the youngest sister, Mary. She was only thirteen years old at the time and usually accompanied Eleanor east. She was not too

Jack Yates around the time of the wedding. (Joiner History Room)

Georgia in her wedding dress. She wore a simple yet elegant ivory-white gown. Her hand rested on the chair to display her wedding ring. (Joiner History Room)

---
[51] George Francis Claycomb, born June 5, 1889.

This close-up view of the Townsend home (see page 32) shows Jack and Georgia standing on the steps. This image was taken in 1889, either just before or soon after their wedding. Georgia's siblings Mary and Fred Townsend stand to the right. Fred had just been elected as a town alderman, beginning a career in local politics that would lead him through three terms as Sycamore's mayor (1895–1899). (Joiner History Room)

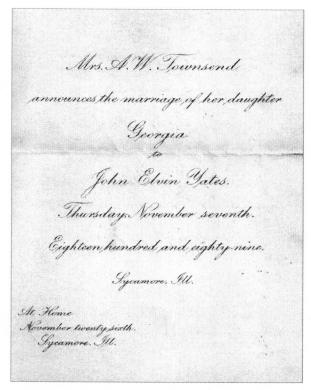

The wedding invitation. (Wendy Jones Smoke Collection)

young to make the trip, so there must have been some other reason as to why she didn't go, possibly related to school, because Georgia and her mother did not return until sometime in September.

Which brings up another question: Why Maine? Eleanor had spent months on the East Coast during previous trips, so it's feasible that she had taken a short jaunt up the Maine coast, fallen in love with the picturesque seaside, and decided to return later for an extended visit. But had Eleanor or Georgia already met Jack Yates before? They could have met on a previous trip to Maine. Or perhaps they met him elsewhere. After all, he spent time in the major port cities of New England—New York, Philadelphia, Boston, Baltimore—all places the Townsend family had visited on previous travels. In July 1888, Eleanor had taken Georgia and Mary on a trip to "the East," but no other details are known.[52] If Georgia had met Jack before—or at least corresponded—was he the reason Eleanor and Georgia traveled to Maine in 1889, and the reason they traveled alone? Instead of a mother-daughter vacation, perhaps Eleanor was accompanying Georgia as a chaperone.

The date of the trip is also significant, because by all accounts it was an ill-timed vacation. As mentioned previously, Anna had just given birth in Sycamore and was in the middle of a difficult recovery, hardly a time for Eleanor and Georgia to leave. In addition, Georgia's grandmother, Ann Denman Townsend, who lived with Georgia and Eleanor, celebrated her eightieth birthday at the Townsend home on August 15. This was a grand and well-attended society event, which included multiple generations of Sycamore's most prominent families. Three of Georgia's four siblings attended, as did most of her cousins, nieces, and nephews. The *True Republican* gave this description: "The parlors were handsomely decorated with flowers, the feast of good things was highly enjoyed, and each and everyone separated with the feeling that it was good to have been there."[53] It is hard to believe that Eleanor would miss such an important family and social event, especially one held in her own home, just to take a pleasure trip with her daughter. So it seems that they traveled to Maine with a specific purpose in mind, one that trumped a new grandchild, a sick daughter, a milestone birthday, and a prominent social gathering.

Whatever the reason for their trip, we can now ask, why Jack Yates? Why a silver-haired sea captain more than twenty years Georgia's senior

[52] *True Republican*, July 11, 1888.
[53] *True Republican*, Aug. 24, 1889.

who would want her to sail with him around the world? From Eleanor's perspective, Jack Yates was more than just a sea captain. As co-owner of his own fleet, he was a wealthy and ambitious businessman worthy of her daughter. From Georgia's perspective, here was a moneyed man who offered independence—from society, from her family, from her home, from everything she has known. Perhaps she sought a life of travel and adventure, something different than the settled lives of her older sisters. All of Georgia's siblings had been born in Illinois and still lived there, while her parents and grandparents had been pioneers who had trekked across the country into an uncertain future and blazed their own trails to success. Perhaps Georgia wanted to escape the confines of Illinois and blaze her own trail, as well. We know that Georgia was well read, and we can glean from her letters that she romanticized the sea (most likely based on her exposure to it in contemporary culture: books, magazines, plays, etc.). Perhaps she saw herself living the adventures she so often read about at home, adventures she wanted to experience before settling down to the domestic life she no doubt knew awaited her.

It is also possible that Georgia may have been under both familial and societal pressure to marry. Georgia's oldest sister, Jennie, married at

Georgia's grandmother Ann Denman Townsend (right) lived with her family in the Townsend home on Somonauk Street. Georgia's sister Mary took this image of her at Christmas in 1899. (Wendy Jones Smoke Collection)

twenty-one. Her older sister Anna married at eighteen. They both wed men they had met in college and already had six children between them. When Georgia met Jack Yates that summer in Maine, she was fast approaching twenty-three, had already left college, and lived at home with her mother, grandmother, and younger sister. By the standards of that time, she was on the fast track to spinsterhood. At church, she was still a member of the Young People's Society. In May 1889, before traveling to Maine, she hosted a social for the society at her home. In October, soon after returning, she continued to take part in their activities. But in March 1890, five months after her marriage to Jack Yates, she hosted her first social for the church's Ladies Society, a group that included her mother. Did her marriage allow her to advance her social status from young person to lady, and could this have added to the pressure to get married in the first place?

Of course, we would be remiss to discount the possibility of true love. Perhaps Georgia and Jack met on the Maine coast, felt an immediate attraction to one another, threw familial and societal concerns to the cool Maine wind, and spent a romantic summer in each other's arms. Perhaps Jack charmed her with his stories of the sea. Perhaps Georgia charmed him with her wit and cultured creativity.

Whatever the case, at some point during Eleanor and Georgia's

Georgia (left) with her friend and classmate Mary Boynton, who would marry Georgia's brother Fred in a lavish ceremony only three months after Georgia's wedding. (Joiner History Room)

summer trip to Maine, they made the acquaintance of Captain John Elvin Yates, struck up a relationship with him, and invited him to visit them in Sycamore, an invitation he readily accepted. We can assume, then, that Jack traveled to Sycamore in October 1889 with the explicit intention of marrying Georgia Wild Townsend.

Next, they appear to have had a hastily arranged wedding, which would account for why this "interesting social event" remained a modest, private affair. We can identify two reasons for the rush. First, Anna Claycomb's health continued to decline. One week after the wedding, Eleanor escorted Anna to Riverside, California, so she could convalesce for the winter. Anna ended up staying in California for six months and returned to Sycamore much restored. Eleanor, however, had to return to Sycamore almost immediately to prepare for another huge social engagement: the marriage of her son, Fred, to Georgia's friend and classmate Mary Boynton, member of another prominent Sycamore family that had made its fortune in finance and real estate.

This wedding was the polar opposite of Jack and Georgia's humble, at-home nuptials. Fred, the oldest Townsend sibling, married Mary Boynton at St. Peter's Church on February 18, 1890—a little over three months after Georgia's wedding. It was the social event of the year, with caterers and musicians brought in from Chicago and several relatives in attendance from back east. The *True Republican* did a large write-up of the event, noting that "everything was done that wealth and taste could suggest to make it an occasion long to be remembered."[54] The scope of the event and the extensive guest list show that the wedding had been in the planning stages for months. A grand wedding for Jack and Georgia may have interfered with (or overshadowed) all of the planning that had already been done.

So Jack and Georgia's wedding may have been rushed by necessity. They had to fit it in before Eleanor and Anna's trip to California and Fred and Mary's grandiose nuptials. Of course, just as we can't rule out true love as the reason for this relationship, we also can't dismiss the idea that Georgia might not have gone in for the spectacle and pageantry of a large wedding. In later years, most of her children married in small, modest ceremonies at the family ranch in Idaho, even though the family was wealthy enough to spring for something much more lavish.

Unless a journal or another set of letters surfaces, we may never know the answer to any of these questions. But no matter the answers,

---

[54] *True Republican*, Feb. 22, 1890.

the end result is the same. Jack and Georgia married. They had several children. By all accounts, they had a good partnership and were deeply affectionate and in love.

Married Life

Jack and Georgia stayed with Eleanor in Sycamore until May 1890, when they departed to make their permanent home in Round Pond, Maine. By that time, Georgia was pregnant with the couple's first child. In November, they began the return journey to Sycamore, spending their first anniversary on the road, and stayed the next few months with Eleanor. Georgia wanted to give birth at her mother's home, where she would be well attended and surrounded by family. It is also possible that she wanted to escape a harsh Maine winter (though northern Illinois had its fair share of harsh winters as well).

Jack and Georgia's first child, Dorothy Yates, was born in Sycamore on Friday, January 30, 1891. The *True Republican* described her as a "thriving baby girl."[55]

Jack had his own motivations for this trip. On March 12, he left behind his wife and infant daughter and embarked on a two-month journey into the nation's hinterlands. The local paper described it as a "sight-seeing trip to Washington and other Pacific coast points."[56] It was actually a trip to purchase land as a future investment. Jack took with him Georgia's uncle, William Henry Townsend, who was two years younger than he. They headed west by rail, spending time in Denver before heading on to the small but rapidly expanding town of Boise, Idaho, which had just been named the capitol of that newly formed state.

At the time of Jack and William's visit, the *Idaho Daily Statesman* declared, "The great spring real estate boom for Boise is on." The newspaper highlighted all the houses and buildings under construction, the real estate being bought and sold, and the large number of land speculators in town. It noted that "the eastern and foreign capitalists, with large sums to place in safe and legitimate enterprises, are looking towards Idaho."[57] The paper might as well have been describing Jack and William, who rode into town with money to invest. Together, they purchased a large tract of land; it is also likely that Jack bought more land just for himself.

---

[55] *True Republican,* Feb. 4, 1891.
[56] *True Republican,* Mar. 14, 1891.
[57] *Idaho Daily Statesman,* Mar. 25, 1891.

After departing Boise, Jack and William headed to Seattle. At that point, William returned to Sycamore and Jack continued down the California coast. On April 28, he registered as a guest at the Baldwin Hotel in San Francisco. Four days later he checked into the Hollenbeck Hotel in Los Angeles. It is possible that Jack visited these cities for more than just land speculation. These were port cities in which the *Willie Reed* regularly did business, so he might have been there to transact further business.[58]

Jack returned to Sycamore on May 18. Two weeks later he took his family back to Maine to prepare for their sea voyage.

Preparing for Departure

For the entirety of Jack and Georgia's relationship, Jack's brother, Oscar, had been away at sea on the *Willie Reed*. He departed New York for Shanghai in June 1889—a month before Georgia's visit to Maine—and arrived in late December. His trip was delayed after the ship ran aground, but it suffered minimal damage. After leaving Shanghai, Oscar sailed back to the United States to load lumber at Port Townsend, Washington. While Jack and Georgia were first settling into their new life together in Maine, Oscar was on his way

Jack and Georgia's first child, Dorothy, was born January 30, 1891, in Sycamore. (Wendy Jones Smoke Collection and Joiner History Room)

---

[58] At the end of the previous year, the *Willie Reed* had been in Chile, setting sail for the United States (port unknown). It could have been in San Francisco at the time of Jack's visit, allowing him a chance to check on his ship and visit with his brother.

to Valparaiso, Chile. When Jack's first child was born, Oscar was in Peru. It is possible that Oscar was in port somewhere on the West Coast when Jack made his journey there and that the two brothers had a chance to visit. As Jack and Georgia returned to Maine with Dorothy, Oscar had the *Willie Reed* in Cuba. From there he sailed to Philadelphia, arriving on June 22, 1891, to discharge his cargo so the ship could be restocked and reloaded for Jack's voyage to Japan. A small setback occurred when the port physician put the ship under quarantine due to reports of yellow fever in Cuba. One of the *Willie Reed's* crewmen showed symptoms, but when his health improved, the ship was allowed to dock and unload its cargo.

Jack, Georgia, and Dorothy spent only a couple months in Round Pond before traveling to Philadelphia. They arrived a few weeks before the *Willie Reed's* scheduled departure so Jack could procure supplies and oversee the ship's loading. One hopes that during their time in Philadelphia, Jack and Georgia were able to meet up with Oscar and Delia, as this would have been Oscar's first opportunity to meet Georgia and his young niece.[59]

The *Willie Reed* departed on September 4, 1891, for a fifteen-month voyage to Japan and back. Jack and Georgia had been married for less than two years and had known each other only a little longer than that. But by all accounts they worked well together, and from Georgia's earliest sea letters we can see that she was excited to be embarking on this adventure of a lifetime.

---

[59] It is unknown whether Delia had accompanied Oscar on this last voyage or whether she had remained in Round Pond.

Part Two: The Letters

The *Willie Reed* by William Pierce Stubbs. Date unknown.
(author's collection)

An Introduction to Georgia's Sea Letters

Seahen. This epithet applied to a captain's wife who sailed with her husband (or any female aboard a ship as companion to the captain or an officer). The ship itself was known as a *hen frigate*. The practice of a captain's wife joining her husband aboard his ship was common in the nineteenth and early twentieth centuries, especially in the whaling industry, where ships could spend years at sea. We learn from Georgia's letters that she met several captains' wives on her journey, including some who were pregnant or accompanied by multiple children. Because of this custom, Georgia had access to female companions when the ship was in port, just as she had access to letters and newspapers from back home. But out on the open sea she had to find the strength to endure isolation as the lone woman living with the *Willie Reed's* crew.

This photograph of Georgia was taken in Galesburg, Illinois, not long before the voyage. Within the first six months at sea, she had lost twenty-eight pounds. (Joiner History Room)

Back in landlocked Illinois, Georgia had enjoyed a comfortable upbringing in a large, privileged household. She grew up surrounded by siblings, hired help, and a never-ending stream of visiting friends and relatives. So her life at sea would have been especially lonely and difficult. On the open water, the ship was a world unto itself, and everyone had a part to play. Georgia struggled with her domestic tasks, admitting early on that she "had never done a washing alone before." But she was quick to adapt and later showed signs of a growing confidence, noting that she did her own ironing and "it looks as nice as anybody's, so there!"

We learn from her first letter that she held romanticized notions about life at sea, almost certainly based on books or periodicals she had read. She was raised in a well-versed family at a time when Robert Louis Stevenson and Mark Twain had become famous for their travel and adventure writing. Many popular magazines published travel journals or features on exotic lands. Influenced by such literature, she began romanticizing the voyage ahead of her. In her first letter, her excitement is palpable as she described new experiences. She was especially enthusiastic when portraying the crew, whom she wanted to use as inspiration for characters in a novel. However, the way she casually referenced this novel—"by the way, did I tell you I was going to write a novel?"—suggests that her plans may have been little more than whimsy. Whether she ever wrote this novel is unknown; she didn't mention it again.

As the voyage progressed, the letters grew shorter, the sentences choppier, the language more stilted; we can feel how her excitement ebbed away as she detailed the tedium of sea life. She emphasized the weather and the wind (or lack thereof). She also dropped the day of the week from the heading of each entry, as though the day no longer mattered. We get the feeling that she was bored, disinterested, tired of the monotony. The wind and weather were all that changed.

She was not only bored, but isolated. Aside from the mates, the cook, and the carpenter, she had little interaction or communication with the twenty-eight-member crew. Not that she sought it. At one point, she confessed that the ship's carpenter was the only "satisfactory person in the entire crew." However, she never addressed the fact that she and Dorothy were the only females aboard the ship or how this situation made her feel.

Early in the voyage, Georgia admitted that for some time she had felt "lower and lower" in her mind. Later she conceded that if she could go ashore, she'd stay there. Any anxieties and concerns she shouldered would have been amplified by traveling with an infant. At one point, when Dorothy took ill, Georgia wrote she "would have given all I possessed for a doctor."

Despite her loneliness and anxiety, Georgia tackled many roles aboard the ship. When a crew member became severely ill, Georgia served as his physician. While the ship was anchored in Japan, Georgia spent days and nights on another ship assisting the captain's wife in nursing the captain back to health. She also undertook other duties as they arose: she filled in for the cook when needed; she mended and knit clothes for Dorothy, Jack, and the crew.

She was clearly learning about life at sea from Jack, absorbing his comments and observations, and then relaying that information with authority in her letters. She was always aware of the ship's location, direction, and speed, information she would have learned (at least at the beginning) from Jack. At one point she casually observed, "We have very little rain, which is very unusual here," and another time she stated that navigating through a specific chain of islands "is often very tedious." Having never been to sea before, she would have known this information only because she heard it from Jack or a crew member. In another example, she shared her husband's frustration: "Here we are, 46 days out and two degrees north of the line – it is enough to drive one distracted." Jack was clearly the one frustrated by this information, because Georgia would have had no frame of reference.

At some point, however, Jack taught her how to use the instruments (compass, sextant, etc.), because she later stated that she was taking readings herself. She took them every day, promptly at 4:00 p.m. This was a common practice for captains' wives. Their husbands would teach them to take sightings, plot the points on the chart, and keep the log books. These would be important skills if something happened to the captain. Stories exist of wives who had to take over ships after their husbands fell ill or died at sea.[1]

Throughout the letters, Georgia provided few specific details on her

---

[1] For more information on women at sea, see Joan Druett, *Hen Frigates: Passion and Peril, Nineteenth-Century Women at Sea* (New York: Touchstone, 1998); Margaret S. Creighton & Lisa Norling, eds., *Iron Men, Wooden Women: Gender and Seafaring in the Atlantic World, 1700–1920* (Baltimore: John Hopkins University Press, 1996); David Cordingly, *Women Sailors and Sailors' Women* (New York: Random House, 2001).

interactions with her husband, so it is unclear how much contact they had on a day-to-day basis. During bad weather or while navigating near islands, Jack got little to no sleep for days at a time and Georgia hardly saw him. But the few times we do get details about their interactions, we can see that they were an affectionate, even playful couple who worked as a team. Jack helped with the wash and taught Georgia how to fish. They had regular conversations and often joked and chided each other. On their anniversary, they made candy together. They relaxed together in a hammock on the ship's deck. They discussed (or debated) child-rearing techniques. When a crewman fell ill, they worked together to try to heal him.

Georgia wasn't the only one gaining new experiences. Jack was embarking into uncharted waters as well. While he had experience sailing with a wife on board, he had never sailed with a child, much less a seven-month-old. And he couldn't look to his brother for advice on such matters, because Oscar and Delia Yates never had children. Jack also had to deal with a wife untested by the sea and naïve to its harsh realities. His first wife, Rockie, grew up on the Maine coast. Several sailors and ships' captains leafed her family tree. She had been seasoned to the sea life, not just by birth, but also through steady acclimation. Her first voyages with Jack were short, sailing back and forth between busy ports in New England. Over the years, as Jack's experience and investments grew, so did the size of his ships and the lengths of his voyages. He began sailing to and from ports in the Caribbean, then South America, and then around Cape Horn and up the West Coast. Finally, with the *Willie Reed*, he went global, sailing to major ports in Australia, Japan, Singapore, Indonesia, and the Philippines. Rockie had sailed with him on the shortest trips, which prepared her for the longest.

Georgia Townsend Yates didn't have that luxury. Knowing nothing of life at sea outside of the romanticized stories she'd picked up from books, she plunged headlong into a fifteen-month adventure for which she was ill prepared. Considering that she was also a first-time mother accustomed to hired help, it is surprising that Georgia made this voyage at all. It is even more surprising that Jack took her.

Knowing all of this backstory, then, it is no wonder that Georgia's first voyage was also her last. Her experience on this long journey may have determined Jack's retirement from the sea, as well.

The Letters

As far as anyone knows, the original handwritten sea-voyage letters of Georgia Townsend Yates no longer exist. They come to us via two different transcriptions, made at different times by different people who were working from the original letters. I refer to these transcriptions as the Sycamore Letters and Seattle Letters, named for the cities in which they were created.[2] The Sycamore Letters cover the entire voyage, while the Seattle Letters are missing several entries from the beginning and end of the voyage. Both sets include different spellings of exotic-sounding islands, cities, and bodies of water. These discrepancies might have resulted from the difficulty of deciphering the handwriting of someone on a ship (at one point Georgia mentioned writing on deck using a lap board while "Dorothy is knocking it about").

The biggest differences are in punctuation and structure. The Seattle Letters contain several run-on sentences that are properly divided in the Sycamore Letters. The Sycamore Letters make heavy use of dashes, while the Seattle Letters do not use them at all. Commas do not always correspond between the sets. The Seattle Letters use numerous semicolons, which are rarely found in the Sycamore Letters. In the Sycamore Letters, longer entries are divided into paragraphs, while in the Seattle Letters many entries have fewer paragraphs or are just a single block of text. The Seattle Letters also use more shortcuts (ampersands, abbreviations, numerals, etc.), while the Sycamore Letters typically spell out everything. The Seattle Letters are also more likely to be missing random words, words that are included in the Sycamore Letters.

Without the original letters, there is no way to know which set most accurately represents Georgia's writing. The letters as they appear in this book follow the formatting of the Sycamore Letters, which are complete, more presentable (aesthetically), and easier to read. But the Seattle Letters proved invaluable for double-checking and correcting dates, ship coordinates, and some confusing wording. Again, it is unknown whether these mistakes were the fault of the transcriber or whether they appeared in the original letters. Whatever the case, the differences between the two sets are minor, the story remains the same, and, most importantly, so does Georgia's voice.

The letters presented here are a complete account of Georgia's voyage.

---
[2] For more information on the different transcriptions and how I came to find and use them, see Appendix C: A Tale of Two Transcripts.

The first entry was dated September 8, 1891, four days after the *Willie Reed* departed Philadelphia. The last entry was dated December 9, 1892, the day before the ship docked on New York's East River. Georgia wrote the letters like a journal, with daily entries, because she could not mail them until she reached a port. So each "letter" she sent home could contain weeks or even months of entries. When she arrived in Japan, she had seven months of entries to send to her mother. While in port, Georgia could send letters as often as she liked and used the opportunity to write other friends and family members.[3] After the ship departed Japan, Georgia faced a much shorter stretch between ports and had more opportunities to send letters home.

While reading Georgia's letters (or any historical letters), it is important to understand the intended audience and how that audience might have influenced the content. In this case, the audience was clearly Georgia's mother, Eleanor Pierce Townsend, as the letters were addressed to her. But Georgia knew that the letters would be shared with the entire family—siblings and their spouses; grandparents; perhaps even aunts, uncles, and cousins—most likely read aloud at her mother's house during the family's large, regular gatherings. This public-style reading of letters was common in this period. Georgia would have taken part in such readings all her life. Her mother would have gathered the children in a sitting room or by the fireplace and read letters from distant friends and relatives: Uncle William on one of his land speculation trips in the West, brother Fred at school in New York, obscure relatives still living in Neversink. The letters would be read again when other friends and relatives came to visit. An example of Georgia's awareness is visible in her letter of December 30, 1891: "Jack and I were talking about you all tonight and wondering where you were and what you were doing." At another point in the voyage, she provided some personal information and asked her mother to keep it private. So she was acutely aware that her letters would be shared with the family.

At this time, local newspapers also commonly reprinted excerpts of personal letters from far-flung residents. The *True Republican* often carried accounts from locals who had traveled or moved west. None of Georgia's letters were reprinted in this way, but the possibility could have lurked in

---

[3] Georgia's letters to other family and friends have been lost (or remain unfound in private collections) save one to her older sister Jennie and one to her younger sister, Mary, which are included in this collection. Georgia's final letter, which covered the last stretch of her voyage home, was addressed to Jennie, because Georgia knew that their mother would set out for New York as soon as they arrived.

the back of her mind while writing.

Another audience she would have considered was her own husband. They shared tight quarters for over a year, so he was bound to see the letters at some point.

And then there was the prospect of the letters surviving for future generations. At one point, Georgia asked her mother to save the letters "as I have not kept any diary, and I shall want it to show to my grandchildren." So she understood that her letters would preserve a record of her trip, a story that would be passed down and shared with a perceived audience that did not yet exist.

A final audience was, of course, herself. During the final leg of the journey, when there were no more stops to mail her letters home, she acknowledged that she wrote the letters because she was lonesome without them. She also stated that she was "so used to having a letter about that I should hardly feel natural without one."

Whatever her thoughts and motivations, a perceived audience always influences content, subconsciously or not. So while reading Georgia's letters, we must consider how Georgia might have shaped her story for her audience, how she might have included or excluded certain details knowing who would be reading her letters, immediately thereafter or in the future.

Voyage Data Sheet

Willie Reed: Three-mast schooner registered to carry 1,500 tons
Dimensions: 200 x 40 x 26 feet
Materials: Oak and yellow pine
Builders: A. R. Reed Company of Waldoboro, Maine
Launched: June 11, 1877, at Waldoboro, Maine
Owners: John E. Yates
Oscar S. Yates
Yates & Porterfield Trading Company
A. R. Reed Company

For this voyage:

Starting point: Philadelphia, Pennsylvania
Destination: Kobe, Hyogo, Japan
Cargo: Kerosene (48,600 cases)
Cleared: Thursday, September 3, 1891
Departed: Friday, September 4, 1891
Number of people on board: thirty-one (and one cat)

John Elvin "Jack" Yates	Captain
Georgia Townsend Yates	Captain's wife
Dorothy Yates	Captain's seven-month-old daughter
Eldridge House	First mate
John Turner	Second mate

Twenty-six crewmen, which included a carpenter, steward, and cook
Crew composed of Norwegians, Germans, Malays, "and a mixture of all nations"[5]
A few of the seamen (as learned from newspaper articles and Georgia's letters):
Alfred Paree (from New Jersey)
Charles Schneider (a German)
George Weltzman
Peterson (a Swede)
Paul (a German)
Jim (an Irishman)
Fisher
Regan

Under the guidance of a pilot, the ship reached the Delaware Breakwater on September 7, 1891.[6]

Georgia Townsend Yates began her letters on September 8, 1891.

---

[4] Because of the clearance date, newspaper accounts reported September 3 as the date of departure. Georgia's reference to the departure indicate that they did not leave until September 4.
[5] *Brooklyn Daily Eagle*, Dec. 11, 1892; *Chicago Daily Tribune*, Dec. 11, 1892.
[6] The breakwater is where Delaware Bay meets the Atlantic Ocean.

Chapter 1: Sailing to Japan
September 8, 1891–March 15, 1892

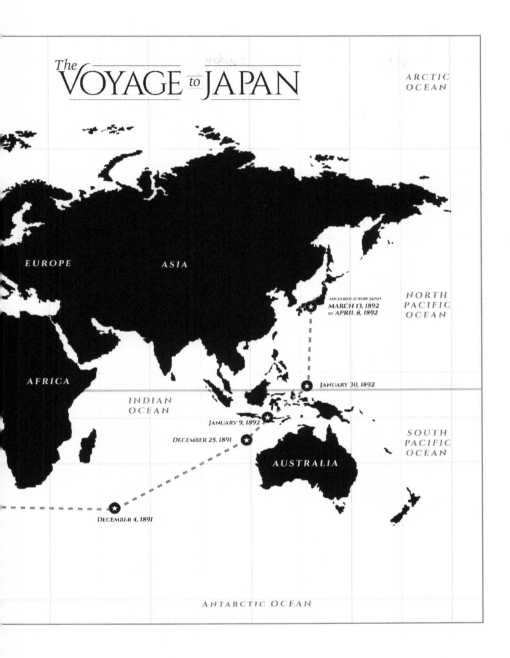

On Board the *Willie Reed*
September 8, 1891[1] [Tuesday][2]

My darling mamma:

Yesterday at noon the pilot left us and we really started for Japan.[3] We had a fine wind all day and made good time. I was sick a little but it did not last long and today I feel about well again. Everything is put away now but I have no doubt I shall have to change them all over more than once. I never really supposed that real, live sailors sang and acted as they do in the books we read, but you would be surprised to see how well they do it. When they hoist the sails or do anything all together they sing such interesting songs – "Whiskey for Your Johnny" is their favorite.[4] One man sings the verses and they all join the refrain while they pull. Then their "Aye, aye, sir" is positively stagey. The mate is steady, quite awfully so, but Jack says he can depend on him if he isn't much of a rusher – he doesn't drink and so he will be some use in port.[5] The second mate I shall put in my novel – by the way, did I tell you I was going to write a novel?[6] He will do beautifully for the brutal mate, – not that he is brutal at all, but he swears as easily as he breathes, and to hear him talk you would think he was longing to kill them all.[7] He is a rusher, if he holds out as he has commenced, but he will be drunk all the time in port. He didn't get sober till we had been out a couple of days.

---

[1] The *Willie Reed* departed from Philadelphia on Friday, September 4, 1891. For the next three days it sailed down the Delaware River and into Delaware Bay, the 720-square-mile body of water between the states of New Jersey and Delaware.

[2] Georgia did not use a standard method of recording the date. In some entries, she began with the day of the week, then month, then date. Later she would switch to month, date, day of the week. She would also go long stretches without including the day of the week. I have chosen to include, in brackets, any information that she left out.

[3] Pilots are responsible for knowing the intricacies of their native ports so they can guide ships safely in to port or out to sea. This saves the ships' captains from having to learn the complexities of every port they sail in and out of. Pilots are used not only for safety, but also for insurance purposes. In this instance, the pilot had steered the ship from Philadelphia, down the Delaware River, and into Delaware Bay. On Monday, September 7, when the ship reached the breakwater—where the bay meets the Atlantic Ocean—the pilot disembarked and took a small, fast ship back to Philadelphia, returning full control of the *Willie Reed* to Jack.

[4] This popular sea shanty is titled "Whiskey Johnny." See Appendix B for sample lyrics to this and another sea shanty mentioned later in Georgia's letters.

[5] This comment implied that it was common knowledge—expected, even—that most of the ship's crew would get drunk in port. A foreshadowing of events to come.

[6] This was the only time Georgia mentioned her plans to write a novel. Outside of these letters, it is unknown if she ever wrote—or attempted to write—about her time on the *Willie Reed.*

[7] This was another bit of eerie foreshadowing or just Georgia's keen observation skills. Jack later described his second mate, John Turner, as "as big a scoundrel as ever trod a deck." (*Chicago Daily Tribune*, Dec. 11, 1892). John Turner did turn out to be a brutal mate. His harsh treatment of the crew lead directly to a brief but violent mutiny.

THE LETTERS

The carpenter I only see as he eats his meals. He is a Swede and Jack thinks him a smart fellow and a good workman. The steward is a short, fat little Jap, very fond of the baby.[8] The cook I only see in the far distance and the crew are, as I said, picturesque. They hang on to the highest place by their eyelids and are as unconcerned as if they were in bed. I am perfectly certain some of them will fall, but I don't say so for fear of being laughed at.

My Lady Dorothy enjoys it all and sleeps and eats and enjoys her salt baths fully.[9] We have no place for her to sleep yet, so she is in bed with me, and Jack sleeps on the sofa. I have spent most of the day on deck in the big chair recuperating after my sea-sickness of yesterday, and if it is as pleasant all the time, I think I shall recuperate all the way to Japan.

This morning we saw a school of porpoise. They jumped clear out of the water and seemed to be having the best kind of a time. Tonight there was a school of blackfish. They were larger, but they are heavy and not nearly so pretty as the porpoise. I am sunburned till my face feels parboiled, but that will not last long. I shall soon have a coat of tan that *no* sun can burn through.

Wednesday, September 9

Today there has been a good deal of wind. It commenced about one last night and Jack has hardly been asleep since. Not that there is any danger, but he wants to know what is going on. I tried to wash some today when it calmed down a little toward noon, but it didn't last long and my clothes are in the tub. I do not yet understand the art of balancing myself and I came near standing on my head in the tub two or three times. I hope you will be able to read this, but the uncertainty of where my pen will strike doesn't help my writing. I have just been on deck, and it is lovely there. The moon and stars are bright and there is a broad path of silver from the port side of the ship to the edge of the world and on the other side, whenever a wave breaks, it shows as plainly as by daylight. We are making good time and I think I'll go to bed, as Dorothy gets me up early in the morning.

---

[8] Throughout her letters, Georgia used the term "Jap" to refer to Japanese people. This term is considered offensive today but would not have been at the time she wrote the letters.
[9] Dorothy Yates was born January 30, 1891, in Sycamore, Illinois. She was seven months old when the ship departed.

Thursday, September 10

A fine wind all day, but from the N.E. We are now four hundred miles out.

The wind has fallen flat tonight and I hope it will start up to the westward when it comes, as that is where Jack says it ought to be. I had thought it would seem to be hundreds of miles to the edge of things and I would be lonesome because it seemed so vast, but it only seems about a quarter of a mile to windward and not over half a mile the other way that you can see, and it seems to be cut square off the edge of a table, and then the water is such a different color out here—in shore, even when it looks blue it is a green-blue, but here it is an indigo blue as though there was too much blueing in it. Yesterday we crossed the Gulf Stream. The water is quite warm though when I attempted to give Dorothy her bath without heating it she protested, and insisted on getting out at once. There is a short, choppy sea here.

Dorothy and I are on deck a good deal. Today I was winding some yarn and had nearly finished my skein when overboard it went, and I had to break it off and leave it there. It has been raining hard but has stopped now. My barrel is full again and I hope tomorrow will be fine so I can get my clothes dry.

Friday, September 11

One week ago today we sailed, and yet in some ways it seems as though it was a month ago, or rather as if I must have gone [on] some voyage before—it is so natural and comfortable. I have been sewing this afternoon, getting some colored clothes under way for Dorothy. I have two flannel shirts nearly finished—I made ten button holes this evening. Isn't that pretty well? The wind is still dead ahead and I think Jack is inclined to swear just a little—I *hope* not, but I *fear!*[10]

Friday, September 18

Friday again! I have not written home for a whole week, but when I have not been doing something else I have been loafing. The wind is still behaving badly – it has not stayed anywhere except in the east for twenty-four hours at a time, and we have only made about one hundred miles a

---

[10] Based on her wording, it is unclear what Georgia feared. Did she fear the unfavorable wind, did she fear what Jack's swearing portended, or was she exhibiting her Victorian sensibility by fearing the swearing itself?

day, but the weather has been perfect most of the time. Saturday I cleaned house a little, and Sunday we all bathed and cleaned off what land there was sticking to us. That evening I put my clothes to soak, but Monday it rained and blew so I let them lie, and I was glad I did, for Tuesday was beautiful and my things really looked pretty well considering I had never done a washing alone before. By Wednesday night I had them all done and I was tired enough to vow I wouldn't wash again until we reached Japan, but I have thought better of it since then. Yesterday and today I have been working on Dorothy's clothes, but when we are on deck I don't progress very fast and it is so lovely I want to stay there most of the time. I think Dorothy has more teeth coming – she works with her mouth all the time and frets to be tended. She grows cunninger every day and she wants to be on deck the whole time. We have been fishing lately. Last night Jack fixed a hook and this morning about ten o'clock we found the stick broken – it is fastened so as to break when the line jerks – but when we pulled it in the hook was gone. Whatever it was took it had bitten off four wires about the size of No.30 thread, so when he fixed it again he used bigger wire, and just after supper tonight we caught a fish a little over four feet long. I don't know what it is but it is a beauty and Jack says it is splendid eating so we'll have something good for breakfast.

Saturday, September 19

It was good, awfully good, and I ate enough to make two people sick, but I think I'll live through it. This morning after breakfast when Dorothy and I went on deck for air, the line was still trailing over the stern and the stick was not broken, but when Jack pulled it in to see if the bait was all right – the bait is a piece of white ray – we found the lower jaw of a fish on it. It hadn't been heavy enough to break the stick and had pulled off – there was a piece of kelp hanging on the hook. I have seen lots of it in the water, but none nearby before – it is very much prettier than that growing on the Maine coast. It is much lighter colored, something like the color of chamois skin, only green tint to it, and it is feathery and fine and the little balls that hold the air are like wax.

After we had the line fixed again I came in to give baby her bath and Jack told the man at the wheel to yell "Fish" if he heard the stick break. I had just put Dorothy in the tub when he called and I dropped her and rushed out to pull it in. Jack came and helped me. It was a Dolphin, about two feet and a half long. I have heard of the colors on a Dolphin lots of times, but I didn't believe much of it – but you have no idea how lovely they are. The back is a bright sky-blue and underneath is cream color; along the sides are spots the size of a dime of bright dark blue,   I   wish you could have seen it as it lay there fairly sparkling in the sun, while the steward hopped up and down in impatience to take it away and dress it.

I wish too you could all have been here this morning to see Dorothy in her bath. I heat it a little just to take the chill off, and she begins to kick the minute she sees it, and such another splashing you never did see. I have to set her tub in the big bathtub to keep from being flooded. She is asleep now on the bed and looking so sweet I have to go and kiss her every little while.

I have been housecleaning this morning and we are all in order now except the bedroom where Dorothy is, and now I must stop writing and go to sewing or I shall never get her clothes finished.

Sunday, September 27

The only reason that I don't write oftener is that I am lazy and don't, but if we don't get on any faster than we do at present there will be quite as much as you will want to read. All the first part of last week the weather was lovely, but there was a good deal of swell that indicated a storm off in the S[outh] W[est] somewhere, but Thursday this was so bad that I sat and held Dorothy all day and didn't try to do anything. Jack was on deck most of the time. By midnight Thursday it was blowing a gale and for forty-eight hours it blew. Friday they put the wooden shutters on so I was in total darkness except for lamps, but the port door of the forward cabin was open and I stayed out there most of the time. Once I was sitting just inside the door watching the waves hop-skipping over the side and across the deck, when along came an extra big one and about two barrels of water came in on me, and I was obliged to change my clothes. It was not nice that day, for everything had to be shut up, so it was close and smelled of kerosene, but the next day – well, I never dressed at all. The only safe place for me was in bed, and there I stayed. When I grew hungry I put a wrapper over my nightgown and went foraging. It was almost impossible to walk – everything was flying in every direction, and to fix things up in the best of order the water came in and the two cabins were about two inches deep. The water didn't get into my room but little, so everything was piled in there, where it rolled from side to side. Dorothy was the best of babies or I should not have known what to do with her. Jack never shut his eyes for over forty-eight hours. He says he never saw a harder gale of wind. There is no romance about a storm like that, and I don't think I ever wanted anything as I did daylight. I wasn't frightened – I was too uncomfortable – and I didn't know this morning but that it was worthwhile to have gone through it, just to enjoy being on deck and realizing what fresh air is, but oh! I wish you could see what a mess this cabin is! Jack has slept all his free time today, and is in bed now. In my tumbles I broke one of the ground glass windows in the after cabin and a lamp. Today has been pleasant and I have been on deck most of the time. We haven't cleared up much, but tomorrow we shall go at that.

Dorothy is as well as ever. She has an upper tooth nearly through and has learned to put her toe in her mouth. She is as good as can be and keeps me employed. One of the sailors has been sick – he had the fever in Rio the last voyage he went and it has left him rather used up. He ought never to have come.[11] We have been doctoring him[12] – I have given him about everything in the medicine chest, and I just have struck it right for he is nearly well again.

I wished you were all here this morning when I was on deck after breakfast in my big chair. The air was so sweet and I was tired enough to be glad to be still. The sea was rolling a good deal yet and the sun was bright. At the other end of the ship the sailors were getting more sail on her and I could just hear "Oh, whiskey made me go to jail, Oh, Whiskey Johnny!" – and then again later – "But I'll drink whiskey while I may, Oh, Whiskey Johnny!"

Saturday, October 3

Just a month since we left the wharf, and it really doesn't seem as if we had been out half that time. If we had only had better wind we should have been across the line before this, but so it goes.[13] The weather is nice enough now, but the winds are very light and not fair at that, but it is very pleasant. Thursday and Friday I did a good-sized washing and ironing, but I have entirely recovered now and have been cleaning house today, so we are neat as wax for Sunday. Dorothy was eight months old Wednesday, and as she grows older she develops an alarming aptitude for mischief. You wouldn't think she could do much with her limited powers of locomotion,

[11] The sick sailor was most likely the man who was reported sick with yellow fever during the *Willie Reed*'s previous voyage, which caused its brief quarantine in Philadelphia (see page 54).
[12] We learn later that she and Jack had been working together to heal the sick crewman.
[13] Georgia always referred to the equator as "the line."

but she is at it from the time I wake up and find her pulling the tufts off of her comfortable until she gives her dinner one last bite and goes to sleep at night.[14] She leads the steward around by the hair and makes herself pleasant generally, but Jack has come to the conclusion that she must cry sometimes. When we began to get more settled he worried because I didn't always take her up when she cried, so at last *he* went to taking her up himself against my advice, so I kept still and let him. She is about as easy to hold as a healthy little eel when she is cross especially, and now he has made up his mind that she must learn that she can't be tended all the time. She is as strong as can be and she has no idea of sitting still on your lap. She climbs and wriggles all the time and is withal so sweet that it's all I can do to keep from eating her.

Wednesday, October 7

We have passed three ships in the last twelve hours. Two were going our way and in the night. The other is just going out of sight now, homeward bound. Jack thinks she is an English iron ship, but she wasn't near enough to signal. We are having fine weather and are in the tropics and trade winds. We make something over two hundred miles a day. Dorothy is having some trouble with her teeth. There are a lot of them that look white and swollen, and I hope they will come through before long.

Thursday, October 8

Yesterday we saw five ships – as many as we have seen altogether since we came out – and today we have seen one, but it was so far off you had to look hard to see it. Last night was beautiful – the stars were so bright and out here you can see so many of them. The Southern Cross has been visible for some nights. The water is not very phosphorescent here, but when a wave breaks it looks as if some stars had become tangled in the foam. The weather is pretty warm – indeed, we have had nothing but warm weather since we came out and I don't expect we shall have much else.

Sunday, October 11

Dorothy is just taking her afternoon nap, and we have had our baths, so Sunday is pretty well through. We are in the doldrums now – the zone of calms between the northeast and the southwest trades. We are something <u>less than</u> seven degrees from the line now but today there is a cool breeze.

[14] "Comfortable" is another term for a soft quilt.

Yesterday it was quite hot. Every night now I walk up and down the deck for an hour. I have not been very well for a week or two, and Jack has made up his mind I do not have enough exercise, so he walks me up and down for that purpose, and I have no doubt it is just what I want. They are taking the ship about and making lots of racket. I hope they won't wake Dorothy. I wish you could see the sea today – so blue and bright and so dazzling where the sun strikes it. I was on deck last night till half-past nine and it was almost as bright as day, though the moon is not half full. The sea on that side was a sea of melted silver, and it was easy to understand how people might become crazy from too much moonlight. I felt rather light-headed myself.

Friday, October 16

Forty days tomorrow since we left the Breakwater, and we are still four degrees north of the line. We have fought our way inch by inch so far and it will be a good six months before we see Japan if this keeps up. We have very little rain, which is very unusual here, but we see plenty of showers going on around us. I have counted as many as six at one time, quite distinct and miles apart, while with us it was the brightest sunshine. It is fine to see a shower coming – I watched one this evening. It did not amount to much when it reached us, but it looked very fierce coming. Dorothy has not been very well for a day or two. I think it is her teeth – she is feverish and doesn't sleep well, but I think tonight she is a good deal better. I have been giving her some medicine. Her fever is about gone and I hope she will sleep better tonight.

Sunday, October 18

Dorothy is all right again. She slept well that night and last night, and nothing can be jollier than she is. She is helping the second mate and the carpenter eat breakfast at present. I can hear them talking to her. We have been making a little more southing during the night and we hope to reach the line by Christmas if we keep on. I have just been feeding Mother Carey's chickens – there is always a flock of them astern. They are lovely little brown things about the size of a swallow, and there they fly all day long and, I presume, all night too. I never saw one *on* the water – sometimes they dip their feet but that is all. I throw bits of bread to them and they scold and fight over it as naturally as humans.[15]

---

[15] Georgia was describing storm petrels, a type of seabird. The name "petrel" derives from St. Peter because of the way the birds appear to walk on water, which Georgia witnessed firsthand. Sailors have ascribed

Tuesday, October 20[16]

Yesterday I washed – I was pretty short of water but otherwise I had no trouble and things looked pretty well considering. For a day or two we have been able to edge toward the south a little, but this morning the wind is dead ahead again. Here we are, 46 days out and two degrees north of the line – it is enough to drive one distracted, and we have about given up hope of being in Japan to celebrate the Fourth.

Friday, October 23

On Wednesday night about midnight we crossed the line. Dorothy and I slept right through it and never felt the bump as we went over. We took the southeast trades soon after and have been going along in a way calculated to raise our spirits ever since. Yesterday about noon we overhauled and signaled a German bark. Jack thinks she was bound for Brazil or thereabouts, and I hope she will report us, so if that storm was along the coast you will know we are safe out of it. She – the bark – was so near that I could see the men in her yards with the glass. I had never felt lonesome a bit till I saw her so near, and *then* I sat there and hardly took my eyes off of her till she fell away astern.

Sunday, October 25

Today I am twenty-five – getting rather old, am I not?[17] It has been beautiful all day, and we have done nothing whatever. The sailors do their washing Sundays and there has been a large wash out forward.

Tuesday, October 27

It is nearly two months now since we came out – it doesn't seem so long, but I should awfully well like to hear from you. Dorothy is getting to be a great girl. She can walk anywhere with her jumper or my hand, and the amount of mischief she can manage *is* remarkable. I am afraid she will never be a model baby, the way she lies down and kicks when she can't have her way looks bad, but she is so sweet and she has so many cunning little tricks it is a shame her grandmothers shouldn't see them. She takes great comfort with the babies in the looking glasses, and then her dolls are a great comfort to her – at present she is eating breakfast with the second

---

several superstitions to storm petrels; some say they are the souls of drowned seaman or cruel captains, while others say they portend bad weather.

[16] This is the first full entry in the Seattle Letters.

[17] Georgia was born October 25, 1866, in Malta, Illinois.

mate and carpenter – I can hear her talking to them – she says mamma and da-da (good-bye) and she *can* say papa but she won't, only just when it pleases her. The old cat and she are very intimate and she is altogether a most remarkable baby. There is a heavy sea rolling today, they say it probably comes from Cape Horn, but anyway it tumbles things about in a most uncomfortable manner. Nothing is safe anywhere unless it is tied, and though I am getting pretty sure-footed I managed to get exercise enough in half an hour last night. Our man that is sick is better again – he was better before and then very much worse again. He had dysentery, the after-effects of yellow fever – there was a while Jack didn't believe he would pull through, but now he is a little better again and we hope it will go on.

Wednesday, October 28

Dorothy has an upper tooth through – if you look back about a month you will see that I said it was almost here. Her teeth seem to worry her some, but she is a very good baby. She will be nine months old Friday and then I expect there will be a row, as I am going to change her meal times to four hours. There is a nasty head sea rolling, so I can hardly walk but I hope it won't last – anyway, if it does, I shall get used to it. Yesterday our latitude was fourteen degrees seven minutes. We are making about three degrees a day now, directly south. The trades will probably take us to twenty-five or thirty degrees and then we'll take the brave west winds.

Saturday, October 31

Dorothy was nine months old yesterday, and a finer baby of that or any other age I am sure there never was. Thursday it rained nearly all day, so yesterday I washed. We have caught so little water that I have only washed what I had to, so I had a big washing yesterday, but Jack helps me and I don't mind it. The trades gave out in twenty degrees and today it is nearly a dead calm, but what wind there is, is fair, a thing we have had very little of since we came out. There isn't a cloud to be seen and the sea is radiantly, brilliantly blue. The air is cool and lovely and altogether life is fair today. Dorothy is outside – I can see her through the window. She thinks I ought to come out and play with her, but I tell her I am writing to Grandmama and she says "da-da" and is reasonable. I have commenced my shawl, and I think it will last for fancy-work until I get home.

Wednesday, November 4

Yesterday I did my ironing and it looks as nice as anybody's, so there! It doesn't always, but this time it does; our invalid is improving – he has begged every time he got well enough to have any appetite, for pancakes – and as his idea of that dainty is flour and water mixed to a batter and fried black in grease, we have not let him have them till yesterday, when the cooks made them for him, but didn't get them to suit, so he went in and cooked them himself. The cook let him, but when the steward found him in there I never in my life heard such swearing, and both the first and second mates can do [a] good deal in that line. The steward is no good – he is dirty and cannot cook, besides being lazy and ugly. We shall get rid of him as soon as we can, but that is one inconvenience of the sea – if your cook can't leave you, neither can you discharge him. We have been having a horrid calm for the last few days, but today there is some wind again. Yesterday I saw as many as fifty flying fish in a school. I have seen a few here and there before, but never so many or so near.

Saturday, November 7

Today is our anniversary.[18] I sat up till midnight last night making myself a present, an awfully sweet silk bag to hold my knitting, and today I made Jack a silk frame for Dorothy's picture, and tonight we made molasses candy. That is, I made it and Jack bothered and wanted to taste all the time. I have blistered my hands pulling it, but it is good. We are having dreadful winds or lack of wind. We are sixteen days from the line and only twenty-eight degrees four minutes. We have had only one day's good sailing since we left the trades, and we shall be lucky if we make the voyage in six months. I am as well as I ever was in my life but I am getting awfully thin – that is, for me – I am quite plump for a skeleton yet. The weather is getting cooler – I have put on Dorothy's thick clothes and my own under flannels. We shall have to have a fire by and bye. We have seen a few Cape pigeons, but have not caught any yet.[19]

Wednesday, November 11

I have been getting lower and lower in my mind for some time. For the last two days we have headed S.W. and N.E. as the nearest we could come to our course. This morning we are going S.E. again, but right on

---

[18] Jack and Georgia married on November 7, 1889. This was their second anniversary.
[19] Cape pigeons, also known as cape petrels, are seabirds common to the southern oceans.

the wind. Dorothy doesn't mind what the wind is, at present she is reading her linen book.[20] I have almost managed to make her understand that it must not be eaten, and she only takes occasional lunches off of it now. I have made her a gray flannel nightgown, and she wore it last night for the first time – it has a drawstring in the bottom so she can't kick her feet out, and her father nearly kills himself laughing at her.

It is bright and sunny but rather cold when the winds are southerly.

7:00 P.M. The wind has been blowing hard all day so that everything has kept collecting on the lee side of the cabin, baby and I included, but we have managed to get along pretty well, and now Dorothy is abed and asleep. I tried to take my usual constitutional and kept it up for some time in spite of the motion, but it began to rain and I gave it up. I fixed a lot of salted almonds today and tonight I am so thirsty I can't drink enough.

November 15 [Sunday]

Sunday again – the time manages to get away with itself pretty rapidly, though I don't always think so at the time. The day after I wrote last it blew a gale and there was a nasty sea running for a day or two after that, but the wind came around westerly, so everything is pleasanter. The weather couldn't be finer, indeed it has been lovely ever since we started, but we can appreciate it now that we have a good wind.

Our sick man is very much worse, Jack doesn't think he can live more than a day or two. He was so much better for a while that we hoped he would get quite well, but he was worse again and we have not been able to do anything to help him. He has eaten nothing for two days, and is so weak he can hardly lift his head. It is terrible – you can have no idea how terrible, to think he is lying there dying and we can do *nothing* for him. If I could go ashore tonight I should stay there – I do not like to be doctor to anyone so sick. Of course Jack has the say about medicine, but he leaves it a good deal to me, or rather we study the thing up all we know together and I mix the doses, but we have had nothing serious except this.

Dorothy keeps perfectly well and grows cunninger every day. She can sit on the floor now in spite of the motion of the ship. She can't creep, but when she wants to go anywhere, she lies down and rolls over and over till she gets there. She likes to walk with me to help her, and she would like to keep

---

[20] Linen books were popular children's picture books in the late 1800s and early 1900s. They were printed on linen because the material was soft and easily washable.

me at it all the time if I would let her. I have made her a flannel nightgown with a string in the bottom so she can't get her feet out, and when she is put to bed she is fastened with three large pins, one in the end of her nightgown to keep her from crawling out and one in the blanket on either side of her because she is put to bed awake and she can lift herself up so far over the side of the bed I was afraid she would fall out. Altogether, if it wasn't for that poor fellow forward, I should be very comfortable and happy.

November 18 [Wednesday]

Day before yesterday they killed a pig, and if you want to know *how* good fresh meat is, just go where you can't get any for two or three months.

It is blowing quite hard today, but in the right direction, so it is all right. We were in 37 degrees S. Lat. and 2 degrees E. Long. yesterday at four o'clock, so we are getting somewhere in spite of all the ill winds we have had. Yesterday I framed that little picture of Mary you gave me in silk, and it's awfully sweet – Jack says so.[21] Today I am making a silk cover for my Bittersweet and when I finish that I am going to frame someone or other in pale pink embroidered with dogwood blossoms. I was going to make it for you, mamma, but I have only a full length picture of you and that isn't a very good one. Please go into Harrison's at once and have your picture taken and send it to me.[22]

Sunday, November 22

37 degrees S. Lat., 14 degrees E. Long. We are flying along at a great rate today. It makes things a little tippy but you don't mind that when the wind is fair and strong. Last Wednesday at noon Fisher died and we buried him about five o'clock. I had thought it would be awful, but it wasn't. I believe I would rather be buried so than in the ground. In all Jack's going to sea, he has never lost a man by sickness before. He has had them washed overboard, but that is different. He did everything he could, he hardly slept for a week, but nothing we could do could help, and he was only twenty-four.

It is colder than it was, but not very cold. We don't have a fire, and Dorothy and I manage to be perfectly comfortable. When I bathe her I shut up the bathroom and light the oil stove to take the chill off. She has

---

[21] Mary was most likely Georgia's little sister, Mary Corey Townsend. It could also have been her sister-in-law and childhood friend, Mary Boynton, who was married to Georgia's brother Fred. There were a few instances in Georgia's letters where she mentioned a Mary, but it is unclear whether she meant her sister or sister-in-law.

[22] Harrison's was a photography studio in Galesburg, Illinois, home of Georgia's sister, Jennie Townsend Webster. Galesburg was also the home of Lombard University, where all of the Townsend children attended college. The youngest sister, Mary, was still a student there at the time of Georgia's voyage.

her fourth tooth through now and I am sorry to say that the upper ones are like mine – large and far apart. I am afraid she is going to look too much like me to ever be a beauty. I have been quite industrious this week – I have framed some of the family in silk. There is no one here to say it for me, so I am obliged to tell you the frames are *very* pretty, also covered a book, made Jack one and a half pair of flannel drawers (the second pair is not finished yet). I had to rip up an old pair of white ones to get a pattern, and I knit pieces for the ankles and they are really quite a success. Then Friday I did a big washing. I am getting so I can do it easier now and the clothes look better. Yesterday I cleaned house generally and altogether I have been busy.

The man who put up the things for the slop chest didn't put in any undershirts or drawers.[23] Jack had a lot of shirts he could spare but only a few pairs of drawers, so I am going to make some. I have some flannel skirts I can rip up – and I am to have the money! If I only had flannel enough I would probably be a bloated bondholder before we reach Japan. Jack expects to have to mortgage the ship to pay my postage if I keep on with this letter.

Thursday, November 26

Lat. 38 degrees 15 minutes, Long. 30 degrees 30 minutes. Thanksgiving Day today, and we spent it far from home, with never a turkey to bless ourselves with. I made a peach custard for supper – I did not dare intrude on the sacred Thursday duff at dinner.[24] We are still moving along at a good rate of speed and still shaking about a good deal in consequence of the same, but we enjoy it none the less.

Dorothy has learned to sit on the floor and even creep in the roughest weather, and if she rolls off unexpectedly she picks herself up and doesn't mind it half as much as I do. She is ten months old Monday and it will be three months since we left Philadelphia a week from tomorrow – it does not seem so long.

November 30 [Monday]

And we are eighty-four days from the Breakwater. Dorothy is ten months old today. She is so sweet and cunning, and has so many pretty tricks – she says "Mamma," "Papa" and "Dada." She will sing if you ask her to, and show me what she has in her mouth when I tell her to, and you have no idea how convenient that last is, because she *always* has something

---

[23] A slop chest is a supply of clothing and other personal items that a ship's crew can access (with a charge deducted from their wages).
[24] A duff is a flour-based pudding that is boiled or steamed in a cloth bag and often contains some kind of fruit (fresh or dried).

in her mouth – usually something that should not be there. She goes to bed wide awake, and hardly ever cries.

She only says "Dada, mamma" in a pitiful little voice when I go out. I have a picture of a cow pinned up, and she admires it very much. She says the cow says "Boo-ah," which is as near as she comes to "moo" and oh! There are so many things she does – I cannot tell you half. At present she and her father are sitting on the floor and she is putting things in his mouth. She thinks this is the next best thing to putting things in her *own*.

We are somewhere about 40 degrees S. Lat. and 45 E. Long. today. We make pretty good time, but we have had two or three calm days lately. We haven't caught an albatross yet. They sail all around the ship and eat everything I throw over, except the hook. They know too much for that – they have had a college education.[25]

December 1 [Tuesday]

The carpenter has just caught an albatross – he measures more than ten feet from tip to tip and looks very much larger than he did flying through the air. He is to be skinned and the skin poisoned[26] – we want two or three treated so and if we can get more we shall use the feathers, which they say are very fine.

December 4 [Friday][27]

Lat. 40 degrees 40 minutes, East Long. 59 degrees 38 minutes – so you see we are getting on. Monday they caught two albatross and Jack and the carpenter skinned them. When we get them home we mean to have them stuffed and I believe Jack intends one for Fred to have in the bank.[28] I tried something new by cooking their livers, and I never ate anything so good. The flesh is too fishy to eat, but those livers – they were as big as my two hands and – well, I wish you had some. Jack laughed when I started, but he ate more than I did.

The wind has been southerly lately, and rather cold, so we have had a

---

[25] An albatross is a large seabird that lives mostly in the southern oceans. Due to Samuel Taylor Coleridge's 1798 poem "The Rime of the Ancient Mariner," many people believed it was bad luck for a sailor to kill or harm an albatross. Sailors did not actually believe this, however, as it was common practice to catch the birds for food and for their feathers.

[26] The skin was treated—or "poisoned"—with a chemical such as arsenic or camphor to prevent bugs or worms from eating it.

[27] The Sycamore Letters have this entry included as part of the December 1 entry. The Seattle Letters have it as a separate entry dated December 4, which is how I have labeled it here.

[28] Fred was Georgia's brother, Frederick Townsend, who was president of Daniel Pierce & Co. (formerly Pierce, Dean & Co.), the bank started by their grandfather.

fire in the cabin. I have been struggling with fate all yesterday and today – I found too late that the stockings I was sure Dorothy could wear when she was two years old were getting too small for her now, so I have been knitting some – or rather, I have just finished the first one. I have nearly worn the yarn out pulling it out and knitting it over to get it right, but really, you know, it looks more like a stocking than I expected, for beside my inexperience I had to knit it of Germantown, and my needles weren't the right size.[29] I made her a new cloak the first of the week – she was ruining that black one by rolling around on the deck.

December 6 [Sunday]

Sunday, and I have worked hard all day, but then, there was no help for it.

Last night the carpenter caught three albatross and a mollihawk, which is much the same kind of bird, only smaller, and today they had to be cared for. One the carpenter skinned for himself, and one that measured eleven feet and over for Jack. The others I had for feathers. I have enough for a small quilt, and the down is *so* lovely I wish I had more, but I may have before I get home. There had been an albatross flying about the ship all day with something tied about his neck. He comes so near that we can see the ends of string or canvass plainly sticking out of the feathers around his neck. I think someone shipwrecked has tied a letter to him, but I can't get Jack interested – he has no imagination.

I have lots of sewing to do this week, and I wish it would rain, for I want to wash, and I haven't a drop of water.

December 16 – Wednesday

Lat. 36 degrees S., Long. 101 degrees E. It is ten days since I have written you – longer than any time since we came, but I have been busy. Last Friday I washed and yesterday I ironed. One day I mended stockings and the other days there has been plenty to do. I have just been up taking my walk. I am not very regular about it, and a good many nights it gets left out altogether.

I have commenced to feed Dorothy this week – she is well as she can be but she has not gained over a pound since we started, and lately she has fretted for her dinner before it was time for it. I don't think sea fare is

---

[29] Georgia was most likely referring to a material purchased from or manufactured in the Germantown area of Philadelphia, popular for its textile industry.

good for milk anyway, so I feed her Mellin's food twice a day and nurse her the rest of the time.[30] I haven't noticed that she is much fatter (I have been feeding her three days), but she doesn't beg for her dinner as she did.

We are going north again now and it is getting warmer. We have not had a fire for several days.

I think a good deal of Fred and Mary nowadays, and wonder how things are going with them and *family*.[31] I can hardly wait to hear. It is nearly Christmas, and I wonder how you will spend it and whether Anna will have the tree she was talking of.[32] I am getting Dorothy's presents ready, and she is to hang up her stocking to the mizzenmast and see if Santa will come here.[33] Blessed baby – she shall have her Christmas.

Sunday, December 20

Lat. 38 degrees S. The winds are awfully light, but the weather is oh! so beautiful. It is warm again now, so that door and windows are open and baby and I are on deck a good deal, and I don't need a coat when I take my constitutional. The sun sets at about half past seven, and I walk from seven till eight – the sunsets are beautiful. You don't have to miss any of it, either, because it can't get behind the trees, and the soft lights and shades last till nearly nine o'clock.

While I was on deck this morning Mr. Turner [the second mate] came aft, evidently looking at something in the water. He said it was the body of a man, the arms stretched out, and it was quite close in. It was gone before I saw it. He may have been mistaken, but it is very possible – someone swept off a ship as men often are, and the body would come to the surface after being in the water awhile.

Jack and I have dressed up today after we took our baths. Jack put on a different suit of clothes and a necktie and I put on a clean waist and a necktie and also curled my hair – *the first time it has been curled since I left Philadelphia*. I have had a sore thumb for a few days – I thought for a while it would be a felon, but it gathered under the nail and is about well now. It kept me awake about all one night and made me pretty cross for a while, but otherwise it did no damage. If we have any kind of wind we

---

[30] Mellin's was a brand of baby food first produced in London in the 1870s. By the 1890s, it had become one of the most popular baby foods in the United States.
[31] Georgia was referring to her brother Fred and his wife, Mary Boynton Townsend. When Georgia departed Sycamore, they were expecting their first child.
[32] Anna was Georgia's older sister, Anna Townsend Claycomb.
[33] The mizzenmast is the third mast from the bow (on ships that have three or more masts).

should be among the islands in another week. I hope we shall have good luck getting through – it is often very tedious.

December 25 [Friday]

Lat. 13 degrees S., Long. 113 degrees E. This is Christmas day, and we are floating in the midst of a dead calm with a cloudless sky. The thermometer stands at 95 degrees. Dorothy thinks Christmas is a great thing. Last night she hung her stocking to the big chair, and this morning there were three bright colored balls and a doll in it, and on the chair two sets of blocks her papa had the carpenter make, a black Dinah doll I made for her,[34] the shell her grandmother Yates sent with her,[35] a tin trumpet and an alligator that goes by clockwork. She just sat back and said "Ga!", but then she went for them. Now she is asleep after her dinner and bath. I have gone back on the feeding and nurse her altogether again. She did not like it very well and it was almost impossible to keep things sweet this hot weather, while there will be plenty more hotter, so I will wait till we are in cooler weather again.

I made Jack some salted almonds and a pretty linen-covered box to put them in, a pen wiper and a toothpick holder filled with some of those superior toothpicks you gave me. I have just been reading my Christmas letter Kate sent me.[36] When it came it seemed an eternity before Christmas, yet here it is, and it has not seemed a long time, either – sixteen weeks today since we left the wharf at Philadelphia.

I wish I could see you all today – how I should like to drop in to dinner with you, wherever you are – but it's well for you I can't – I should talk you all deaf and dumb. You would not have a chance to get a word in edgewise. Think of not having a woman to speak to in four months.

I have given Neptune my Christmas greeting to take to you – he is to telegraph it from New York.

December 30 [Wednesday]

Lat. 11 degrees 15 minutes. It is *hot!* The thermometer has stood between 100 and 105 degrees all day today. We have had almost no wind

---

[34] A Black Dinah doll was a doll of an African-American female attired as a house servant. By modern standards these dolls would be considered offensive, but they were quite common in Georgia's time.
[35] Grandmother Yates was Jack's mother, Sophia (Blunt) Yates, who lived in Round Pond, Maine.
[36] Kate was most likely Georgia's aunt, Kate Brundage. Kate was the sister of Georgia's father. She was married to Frederick Brundage, a first cousin of Georgia's mother. They lived in Malta, Illinois, just south of the farm where Georgia was born. The two families were close and paid regular visits to each other's homes.

until tonight, and we have had the full force of the sun. Dorothy is not well – I noticed it first Christmas day, but she is cutting teeth and that, with the warm weather, I think account for it, but of course I am anxious. She is all broken out with a rash and is pretty fretful. She has one more tooth through, and that an upper one – that is why it surprised me – I was looking for lower ones.

We have had very calm weather lately, and are only making about a degree a day. There is a pretty good wind tonight, but there is no telling how long it will last. It doesn't rain, and I can't wash. We have had so little rain the entire voyage that we are being awfully careful of the water. Jack and I were talking about you all tonight and wondering where you were and what you were doing, but when we happened to think of the coal fires you were sitting by it made us so warm we had to stop and go fan ourselves. Jack shot a turtle yesterday, but he didn't come up near enough so we could get him. It is thundering and lightening at a great rate tonight, and I can hear the sailors forward pouring water over one another. I wish someone would pour some over me.

January 5, 1892 [Tuesday]

Lat. 9 degrees 14 minutes S., Long. 115 degrees 30 minutes E. and here we have been for five days and here we are likely to stay for fifty, so far as I can see. The sea looks like a piece of blue-tinted mother of pearl. There is not a ripple. We are about thirty miles south of Bally Island or Lombok Strait, and the land has been visible at times.[37] Yesterday we could see seven mountain peaks, but now the clouds hang about there so we can see nothing. We have had two rain squalls and caught a lot of water, so we are all right in that direction again. I did an immense washing Saturday, but it is still in the basket as it is too hot to iron. The thermometer has not been below 100 degrees this year.

Dorothy is about well again and her rash is nearly gone. I think the fresh instead of salt water for her baths is best, this hot weather, but she has two great boils on her head and has had five before. She is awfully good but I am afraid I shall have a time with her when it gets cooler and she is not tended so much. I have made Jack a pair of pajamas which are nothing more or less than Mother Hubbard pants that they wear in these climates. He told me how to make them, but I had no pattern, and they are peculiar, to say the

---

[37] They were nearing the Indonesian island of Bali. Lombok Strait runs between the islands of Bali and Lombok. It separates the Indian Ocean and Java Sea.

least.³⁸ We have an awning up aft here, and that helps, but it is pretty warm. I think we had better have a meeting to pray for wind.

January 10 [Sunday]

About halfway through the Straits of Allas.³⁹ We managed to get wind enough to get in here at last, and Friday afternoon about four o'clock we cast anchor off Labuan Hadji, and for the first time since we left Philadelphia we stood still.⁴⁰ The town was in plain sight, a long line of huts along the shore, and we looked anxiously to see if there wouldn't be a boat off with fruit. At last about dark we saw one coming and were glad, but we were glad too soon, for when they came alongside they had not a thing. They wanted to buy a musket, and when we wanted to know why they didn't bring fruit they said, "Tomorrow."⁴¹ As we meant to sail at daylight that didn't do us much good. One man knew a few words of English.

But it so happened that it was either a headwind or a dead calm all day yesterday, so when they came we were still there, but their stuff was very little good. But it was fun to watch them. All they wore was a kind of skirt made of a piece of cloth wrapped around them and a scarf around and around their waists. Jack got a lot of sweet potatoes, some chickens and a lot of eggs, but all my pleasant memories I expected to have were knocked on the head by what happened afterwards. When I am home again I may remember that part and then I will tell you about it, but at present the rest of the day is all I can think of. Among other things, they [the crew] brought rum aboard – we don't know when or how – but at noon there was a row forward and a lot of them were crazy drunk. Then they began coming aft, and Jack and the mates did everything they could

---

³⁸Mother Hubbard pants are loose-fitting, ankle-length trousers similar in design to harem pants or pants worn by French Zouave soldiers.

³⁹ The Alas Strait runs between the Indonesian islands of Lombok and Sumbawa. Because Georgia provided little description of the setting, I include the following excerpt from the sailing journal of Lady Annie Brassie, which provided a brief description of her trip through the Alas Strait in 1887:

> Friday, April 22nd—The Strait of Allas is one of several navigable channels by which ships can pass from the confined waters of the Eastern Archipelago into the Indian Ocean. It divides the island of Sumbawa, famous for possessing the most active volcano in the world, from the island of Lombok. At the eastern end of Lombok, a magnificent peak rises to a height of 12,000 feet, and overshadows the narrow channel beneath with its imposing mass…. The peak of Lombok stood forth clear of cloud, in all its majesty and grandeur, backed by the glorious colours of the evening sky.

Annie Brassey, *The Last Voyage* (London: Longman's, Green, & Co., 1889), 217.

⁴⁰ Labuhan Haji is a port city on the east coast of Lombok.

⁴¹ The native traders likely wanted a musket because eastern Lombok had recently launched a rebellion against the island's Balinese ruler. The rebellion started around the time the *Willie Reed* departed Philadelphia, so it is unknown whether Georgia or anyone else on the ship knew of it. The rebellion lasted until 1894, when the Dutch intervened and made the island part of the Dutch East Indies.

to get them to go forward and keep quiet. They accused the second mate [John Turner] of everything under the sun, but he kept still and did not do a thing to make it worse. They [Jack and John Turner] put two of them in irons two or three times, and then they would say they would go forward, and they would take them [the irons] off, and at last nearly the whole crew were aft here saying what they would and would not do. Jack got his pistol and gave Mr. Turner the other and Mr. House [the first mate] the rifle, for it was mutiny, and there was nothing else to do. They tried to get them [the crew] forward without firing, but they kept at Mr. Turner till he fired, and I wonder he waited so long. That started them – one of the ringleaders was hit in the leg, and the other ran through the cabin out aft, and they fired at him but the bullet hit another man. They went forward after that, but I was nearly wild, and I shan't know a minute's peace till we get rid of them. They are quiet as mice today, but they can't be trusted a minute – I am afraid too they will have more trouble with Regan, the ringleader who was not shot. Today we started on again, but I don't think we went ten miles, and now we are at anchor, but we hope to start up with the tide in the morning.

The Mutiny on the *Willie Reed*

At this point, a brief interlude is required to extrapolate on the events of Georgia's previous entry. What Georgia just described was, in fact, a mutiny. For an event that made her "nearly wild" and would have set the tone for the rest of the voyage to Japan, Georgia's details were surprisingly sparse. Additionally, she barely mentioned the incident again in the rest of her letters. Perhaps she was trying to spare her family from any further anxiety. It was also possible that she didn't know many of the details herself. After all, the mutiny happened quickly and we can't know for sure how much Georgia witnessed. Or Jack might have instructed her not to write much about it, knowing that it could become a complicated legal matter.

After the ship's return to New York in December 1892, however, several accounts of the mutiny appeared in newspapers across the country. These articles, which ran under sensational headlines such as "Mutiny and Bloodshed," "Fight with a Mutinous Crew," and "A Shooting Affair at Sea," included differing eyewitness accounts that help fill in the details missing from Georgia's letters (though some of these details are either biased or suspect). The articles also provide some backstory to show what incited the mutiny in the first place, details that Georgia may not have known. It is interesting to note that the man Georgia mentions as the supposed ringleader, Regan, was never named in any of the newspaper articles.[42]

The following account is taken from an article titled "Mutiny and Bloodshed" that ran in the *Brooklyn Daily Eagle* on December 11, 1892:

> Everything went along smoothly for the first few weeks and then Turner began to tell him [Captain Yates] that the men were grumbling about their rations. He paid no attention to the remarks because, he claims, the men were well fed.[43] It was not long, however, before he found that Turner had supplied the men with duplicate keys to the ship's larder and that they were stealing the ship's provisions. Two days before arriving at the Sunda Straits Turner forsook the men and took the duplicate keys away from them.[44] This aroused ill feeling.

---

[42] It is possible that the letter transcribers misread Georgia's handwriting, but the name was written as "Regan" in both the Sycamore and Seattle Letters.

[43] An article in the *New York Press* (dated December 11, 1892) defended Jack's decision to ignore the complaints, noting that he had "followed the sea for many years, and knew the men were receiving as good rations as could be had on any sailing vessel."

[44] All of the newspaper accounts reported the mutiny taking place in the Sunda Strait, which runs between

When the ship arrived at the straits several boat loads of Malays came alongside and sold the sailors a large quantity of fruit and whiskey. The result was that the sailors got drunk and several of them, Alfred Paree, George Weltzman, Charles Schneider, a Swede named Peterson, and an Irishman called Jim, became very noisy.[45] Paree, who seemed to be the leader, created so much disturbance that he was placed in irons. The following day the crew refused to work unless Paree was released. The captain acceded to their demands. The next day the captain, who was in the cabin, heard a racket on the deck [and upon] going out found that the man Paree and Turner were having a war of words. Paree accused Turner of stealing the keys and food for them. Turner replied with vile epithets and Paree made a rush for him. As he did so Turner pulled his pistol and shot Paree through the body. He then fired three other shots promiscuously, inflicting flesh wounds on Schneider and Peterson. By the time he had ceased firing the sailors had all disappeared in the forecastle.

From that time until the ship reached Kobe, on March 16, 1892, the captain said everything seemed to go all right, but the low mutterings of the crew whenever Turner gave an order made him live in constant fear that the men might be goaded into another outbreak.

Several newspapers also ran a version of the story in Jack's own words. The most complete version ran under the headline "A Mutiny at Sea," in the December 12, 1892, issue of the *Dallas Morning News:*

> The *Willie Reed* left Philadelphia early in September, 1891 for Kobe, Japan. I found it impossible to get a crew that I desired and so had to take what I could get. The crew was composed of Norwegians, Germans, Italians and a mixture of all nations.[46] Everything worked smoothly for some time. Then the men grumbled over their food. Later I discovered that the crew were stealing food from the steward. My second mate, John Turner, before the voyage was over, I found out was as big a scoundrel as ever trod a deck. He instigated the men in their robbery of the ship's stores and encouraged them in their misdeeds.
>
> Shortly before reaching Kobe, March 16, when at the Straits of Sunda, a number of Malay Indians came on board and brought some

---

the Indonesian islands of Sumatra and Java. The *Willie Reed* would later return by this route, but from Georgia's letters we learn that at the time of the mutiny the ship was in the Alas Strait, anchored off the port village of Labuhan Haji. The details in Georgia's letters make it unlikely that she was wrong about the location, but the newspaper stories were taken directly from Jack's account. It seems odd that he would have been so mistaken. Also, it is unclear whether Jack ordered Turner to collect the keys or whether Turner did it for his own reasons.

[45] In some accounts, "Paree" was spelled "Pare."

[46] In all other newspapers that carried this story, this line stated that the crew was composed of "Norwegians, Germans, Malays." For reasons unknown, the *Dallas Morning News* was the only paper to change "Malays" to "Italians."

whiskey which they gave to the crew. The crew got drunk and refused to obey orders. They said they would like to see the man who could make them obey. Only a part of the crew, about eight men, seemed to be actually mutinous. One of them, Alfred Pare, I considered to be the ringleader, and I thought it time to take some action. I ordered him placed in irons. Some of the men were still faithful in carrying out my orders. Pare's companions were too drunk to offer much resistance. The next day all of his companions came to me and demanded Pare's release. They refused to work the ship unless their demands were acceded to. I saw that I was surrounded by a gang of bold and desperate men who would hesitate at nothing. Their conduct was mutinous, but there were no war vessels or police about from whom I could get assistance. I therefore released the man.

Shortly after I heard a great noise on deck. I opened the door of my cabin, and looking out saw the second mate, Turner, with the turbulent part of the crew surrounding him. Some of them were shaking their fists at the mate. I heard Pare say to Turner, "You are as big a thief as us," and at the same time Pare made a jump for the second mate. Turner pulled out his pistol and fired. The first shot stopped Pare. Turner fired again and wounded a sailor named Schneider. His third bullet wounded another of the crew named Peterson. Then the mutineers scattered and fled to the forecastle. They had no arms and were cowed.

Another version of events was told by Charles Schneider, the German seaman whom Turner shot during the incident. He claimed that the crew's discontent grew out of conditions in which they were "practically starved, compelled to work at unreasonable hours and subjected to brutal treatment."[47] His version of events ran in the October 20, 1892, issue of the Virginia newspaper the *Alexandria Gazette:*

When Captain Yates was shipping his crew in Philadelphia, he took special pains to engage only foreign seamen. One man shipped under an assumed name in order that he might be engaged. When we left Philadelphia our relations were very friendly, but the further we got from America the scarcer became the grub. It finally got so bad that we got only salt meat and bread, and very little of that. At the Straits we were forced to lie to on account of a storm.

Three of the crew got drunk. They were Alfred Paree, a Jerseyman who shipped under the name of Nieuwenhuizen; an Irishman known as Jim, and George Weltzman. Paree's grievances magnified themselves, and he went to Captain Yates and complained. The captain put him in irons. We all refused to work unless Paree was released. The captain finally let him off, and while Paree was working, John Turner, the mate,

---

[47] *Evening World* (New York, NY), Oct. 19, 1892.

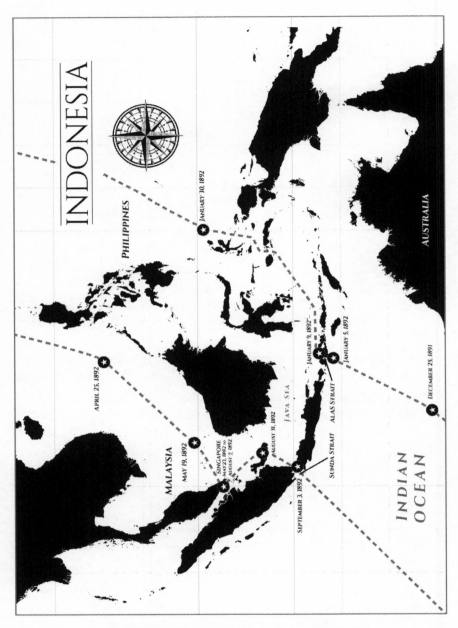

The *Willie Reed's* path through the Indonesian islands

approached him and began to curse him.

Paree jumped for Turner. Captain Yates heard the noise and came out of the cabin with a revolver in his hand. Turner grabbed the revolver from Captain Yates and shot Paree. Turner kept on firing the revolver and one of the bullets struck me in the groin.

This same account also ran in the Texas newspaper *The Velasco Times*, which included the additional information that the shooting took place just as the crew was getting the ship ready to depart and that Peterson was struck by the bullet that passed through Schneider.[48] The New York newspaper *The Evening Times* speculated that if Schneider's story were true, "it will very likely make trouble for Capt. Yates and his second mate, John Turner."[49]

Most accounts mentioned a bad storm just before the mutiny, implying that this storm contributed to the men's behavior. This was likely the "two rain squalls" mentioned in Georgia's January 5 letter. Georgia's recollection of the storm shows how her life on board the ship was vastly disconnected from those of the sailors. While they struggled on deck in miserable conditions, she was delighted that she was able to catch an abundant amount of fresh water so she could do the wash and properly bathe Dorothy.

A key detail missing from the newspapers was the heat. Georgia mentioned that the temperature had hovered above one hundred degrees for days. So before the mutinous crewmen had consumed any whiskey, they were already hot, hungry, dehydrated, and miserable.

There is no record of Jack or his second mate being charged or punished for their actions. Whatever the true story behind the mutiny may be, it is certain that neither Jack nor Georgia could rest easy until they reached Japan, a state of affairs that Georgia barely mentioned. Jack noted, "Until we reached Kobe I kept constant watch, for I was afraid that the mutineers would do something desperate, especially as Turner was as great a scamp as any of them."[50]

---

[48] *Velasco* (TX) *Times*, Oct. 28, 1892.
[49] *Evening World*, Oct. 19, 1892.
[50] *Dallas Morning News*, Dec. 12, 1892.

January 17 [Sunday]

Sunday, and we are sailing along with a fine breeze. I hope it is the beginning of a pleasanter week than the last. Monday night three of the men slipped over the side and started for the shore. Two of them we don't know what became of, they may have reached the shore, but it was about two miles and a strong current. The third came back. Jack had just laid down for a nap, and the mates didn't keep watch properly. It has left the ship awfully short-handed, one man dead, two laid up and two gone. The two hurt are doing well and will be about again after a while.[51]

We creaked out Tuesday night, but all the week until yesterday it has been *calm* – Oh! so calm. We should not have made an inch if there had not been a current going our way, but yesterday there was some wind and today there is a good breeze, but rather ahead. Jack is having rather hard work now, but there is no help for it – there are so many islands and reefs through here. We passed an island today that has a live volcano on it – it is the first live one I have seen, though there have been plenty of extinct ones. Dorothy has been far from well – the weather has been very hot and between that and her teeth, her bowels have been pretty bad, but the tooth is through and the weather is cooler and I think she is a good deal better. It will be nineteen weeks tomorrow since we left the Breakwater, and if we have good luck we shall reach Japan soon and I shall hear from my dear people again. I have enjoyed many things, and I am not sorry that I came, but no one can know how I want to hear from you all, to know that you are all well, and how Mary and the new baby are.[52] Jack and I were saying last evening that it was about time for Uncle George to come over to see if Nellie was back from the city all right and if they were taking proper care of the baby.[53]

January 24 [Sunday]

Sunday again and we are still fighting our way among the islands. There is plenty of wind but it is dead ahead so that while we sail between one hundred and fifty and two hundred miles a day, we made yesterday,

---

[51] In Charles Schneider's newspaper account of the mutiny, he identified the three men as "[George] Weltzman, a German named Paul, and a Swede." He noted that they had jumped into shark-infested waters and that Paul returned to the ship due to cramps, but the other two men were never seen again. He also implied that the men deserted to escape the brutal treatment they received under John Turner, but it was also likely that they were trying to escape the punishment they undoubtedly faced when they arrived at Japan.

[52] Georgia's nephew, Charles Boynton Townsend, was born January 1, 1892, at the Townsend residence on Somonauk Street. He was the first child of Fred and Mary Townsend.

[53] Uncle George was George P. Wild, who was married to Georgia's aunt Sarah (Pierce) Wild. Georgia was named after him. Nellie was his daughter, Elinor Wild, who was named after Georgia's mother.

twelve, and the day before seventeen miles on our course – and one day's fair wind would take us into open water! Jack hasn't been to bed in a week, only getting such cat naps as he can in the chair or the hammock on deck, and awfully little of that.[54]

Dear daughter Dorothy will be a year old on Saturday next, and is quite the smartest and cutest baby that ever was. She can all but walk alone, and talks everything but words, which she doesn't need as she can make anyone understand perfectly without. She is inclined to be a trifle cross just now on account of her teeth but it doesn't amount to much. At present she is on the floor playing, but she wishes to go on deck. As I show no sign of taking her, she first cries "mamma, mamma" and then leans over and bumps her head against the wall, being very careful not to bump it too hard. It doesn't seem possible that she is a year old. We are to have the flags set on her birthday. Jack's is five days later, and then we shall each have had one on board the *Willie Reed*. Her ladyship has now wandered in and stands beside me, so I think she wishes to write to grandma a little.

Dear Grandma, I am pretty well and I send you my love and a kiss. Here it is X. Your loving granddaughter, Dorothy Yates.[55]

There is a real kiss there, and she held the pencil all the way, and is sending more kisses all the time than I can write.

January 30 [Saturday]

My lady Dorothy is one year old today, and the flag is set in her honor, and she has also received her birthday presents and had on a white dress. She is now in bed and asleep just as if she was only eleven months and twenty-nine days old. Yesterday we left the last island of Gillolo Passage and Jack was able to go to bed last night – the first time he has really turned in in nearly three weeks – and I think he enjoyed it.[56] The wind is very light, but we have started on the last part of the voyage, and we can see Japan on the chart with only about 34 degrees of the Pacific between it and us, and that is a very short distance considering what there is behind us. We have the most beautiful tropical sunsets, and I swing in the hammock and watch them and wish that all you dear people were here to enjoy it with

---

[54] Georgia's letters implied that Jack's lack of sleep was due to the difficulty of navigating through the islands, but we learn from the newspaper accounts of the mutiny that he also lived in fear of a second uprising due to the volatile and extremely tense climate aboard the ship.
[55] The transcriber of the Sycamore Letters made a note that this line was written in pencil.
[56] Gillolo Passage runs east of Gillolo Island (modern-day Halmahera), one of Indonesia's northernmost islands.

me, for it's the sort of thing that *no* description is equal to, and I was never very much at description anyway.

You have no idea how *many* excitements we have – a school of porpoises or blackfish or sometimes a whale – then there are squalls, and I think I have seen as many as a dozen waterspouts while our own little world aboard here furnishes endless variety; however, I am sorry to say unpleasant variety – in fact, I think the carpenter is the one satisfactory person in the entire crew, and that means everybody aboard except Jack, Dorothy and me.[57] I shall get at my letters soon – I have so many to write, and I can tack on a postscript after we get in.

February 2, Tuesday

Yesterday I washed, and as it is over three weeks since I washed everything before, there was an awful lot to do this time. I am waiting for my irons to get hot so that I can iron, but the steward can't see to make the fire burn. I think he must have come to sea for his health and to rest – anyway, he won't work, and as he is not ornamental he is about as useless as he could be – but we shan't have to put up with him long, and that is a great comfort. We are nearly eaten up by mosquitoes just now, the wind is light and doesn't blow them away, and they are as thick as flies in August. Day after tomorrow is Jack's birthday and then we shall each have had one since we left home.

February 4, Thursday

Jack's birthday today.[58] Winds very light – they were better yesterday, but this morning nowhere again. There is a lovely moon now – Jack and I were out in the hammock last night and it was almost as light as day. I am weaning Dorothy and she doesn't like it very well but I think she is getting fatter.[59] She has lost a good deal of flesh this hot weather.

February 6, Saturday

I hope we shall get to Japan by the first of May, but it's doubtful if this wind holds. It is dead ahead, and awfully light at that. We have been going southeast for twenty-four hours now and the wind shows no sign of changing.

---

[57] While Georgia was almost certainly alluding to the mutiny, it was possible that other "unpleasant" incidents took place that she chose not to write about. Even if nothing specific happened beyond the initial mutiny, the ship's atmosphere was undoubtedly stressful.
[58] Jack was born February 4, 1845. On this day, he turned forty-seven years old.
[59] Georgia could have been preparing to have another child, and she might have been hinting at that by telling her mother that she was weaning Dorothy.

THE LETTERS

February 10, Wednesday

The wind is still ahead and the current dead against us; it is slow work and very trying on the nerves. Today is your birthday, dear mamma, and I hope it will be a happy one, with many more to follow, for whatever could we do without our darling mother?[60] Dorothy has sent lots of kisses to you by the way of the man in the moon. She has just made her first astronomical discovery – she has found the moon and I have told her about the man that lives there. She is greatly interested. She is trying hard to walk, but the ship has rolled a good deal for a day or two and she has to hold on to things pretty tight. I do myself.

February 12 [Friday]

The trade winds have us firmly now, so we know what to expect for a while at least. We can't make our course, but we don't come very far from it, and the wind is strong. We have had smooth weather so long that I have to begin all over again about leaving things where they will slide off. One of the men fell overboard the other day while it was smooth, and they had him in again in no time. If he went over today, I am afraid there would be small hope for him, but they look out for themselves when it's rough. We are in 136 degrees 46 minutes E. Long. and 7 degrees 15 minutes N. Lat., making a west by north course, but the current here keeps us about north. Hiogo is in about the same Long. and about 34 degrees 40 minutes Lat. and that means three or four weeks yet at least. I am afraid we shall go over our six months, we have had such hard luck around here.

February 13 [Saturday]

We moved about two hundred miles in the direction [of] Japan yesterday, and I begin to hope again.

February 14 [Sunday]

We still move, and I still hope. Today is St. Valentine's Day – I did not think of it until this morning, or I would have made Dorothy one. Dorothy has had her bath and Jack is taking his now, then I come on and we are all dressed for Sunday. Yesterday Jack got my trunk out and I overhauled my clothes, and most of them I put back again as being useless or needless at any rate. When I tried them on I began to see that I had lost flesh, as I have taken up some two inches

[60] Eleanor was born February 10, 1839. On this day, she turned fifty-three years old.

or more in the seams. I am going to be weighed tomorrow.

February 16 [Tuesday]

I *was* weighed, and I do not wonder that my clothes did not fit. I weigh 124 pounds – think of it – 28 pounds have I lost since I left home, and I was never in better health. Jack has lost a good deal too, and Dorothy only weighs 18 pounds, but I guess we shall all pull through. I am way down in my mind again – the wind has twisted and we are going off E. by N. We shall be another month if we have to tack all the way there.

February 18 [Thursday]

Fred and Mary's anniversary today.[61] I wonder, do they celebrate? We have had all kinds of weather since I wrote last – twenty-four hours it blew nearly a gale and there was an awful sea running from ahead. Today it is quieter and we have been heading pretty well up, so my mind is quite calm. We must be nearly up to 20 degrees now, and there are *only* 15 degrees ahead – that is, straight ahead, but there is no knowing how far we may have to go to cover it. Dorothy is at last the possessor of a monkey. Jones has managed to give it to us in spite of us, and I fully expect we shall rue the day, however.

February 21, Sunday

We have left the trades again, or they have left us. Yesterday was nearly calm all day, but today we have had a good breeze again. We are housecleaning with a vengeance, but I don't have to do any heavy work. We have one of the sailors to rub the woodwork and he is doing it beautifully. We didn't dare trust the steward, he would slap over and not half do it. I am going to wash tomorrow if it is a fine day.

February 24, Wednesday

It has been rainy all the week, so I have not been able to wash yet, but I shall tomorrow unless it rains pitchforks, for I need some clean clothes for Dorothy, and besides, oh! joy – we near our destination, and my lady baby must have some white clothes to wear. Her teeth are bothering her a good deal, and it makes her fretful. She wants one to entertain her all the time – the carpenter and I have been making a cot for baby. It is on the lounge and consists of a board at the side and one at each end of that across to keep her

---

[61] Fred and Mary Boynton Townsend were married on February 18, 1890. This was their second anniversary.

from going up or down. She has always slept with me and Jack in the spare room, but she doesn't sleep very sound and if I am restless and move any it wakes her, and I think it is better for her to sleep alone anyway.

I have never found a moth until yesterday – my trunks had not a one, but in turning out the things in Dorothy's drawer, some of her flannels were eaten a little. Her two commonest blankets were spoiled and one or two other things were a little eaten, but nothing badly. I had not had camphor in there as I used those things all the time, but I have put plenty in now. Only six degrees more if we only have good luck – we could do it in two days easily, but it will probably be three or four.

February 26, Friday

Three or four days, indeed! It will probably be twenty – yesterday was a dead calm all day and up to noon today, then there came a gale of wind *dead ahead*. It has gone down some, but not much – fate is hard, hard! Dorothy went and got her monkey last night, and we have him on our hands now. He slept in the wheelhouse last night, but he was so cold and miserable this morning I brought him down here, and he is in his box in the bathroom. He does not like to be alone, and cries a good deal, but I hope he will get over that. I gave him a bath today and he didn't like it at all. He has taken a great liking to me and wants [me] to tend him all the time. Dorothy loves him fondly and he likes her very well – he is also very fond of soup.

Dorothy is getting a little flesh on her bones. When we weighed her Monday she had gained a pound and an ounce since the last Monday, and I can see that she looks fatter. She is so cunning, I wish you could see her. If she does anything naughty she begins to kiss you as hard as she can so you won't mind. I have a lot of the family photos stuck up in the looking glass, and she will stand and look at them and point her finger and squeal until I want to pick her up and carry her right home to show to you all. I thought this quire would carry me through, but I shall have to start another.[62] I really do pity you all if you try to read this, but you needn't do it all at once. But whatever you do, save it for me, as I have not kept any diary, and I shall want it to show to my grandchildren.

February 28, Sunday

The old ship is first standing on the end of her bowsprit and then on

---

[62] A quire is several sheets of similar-sized paper collected in a book or inside a cover. This was what Georgia was writing in.

her rudder, and everything else is going likewise. Dorothy's monkey is a pest – I said before we got him that I doubted the wisdom of that act, and I doubt it very much more now. He is little, but oh!! my!! – the mischief he can find to do, even in the small scope of his string. One day he untied the string that holds the foghorn on its shelf, and that fell off and knocked a big dent in it and all the time he reaches way beyond where I have any idea he can and pulls down towels and then he cries a good deal, but Dorothy loves him and so it is all right.

The wind is behaving awfully. Four days ago I hoped that we might be safely anchored by this time – Jack says *he* never had any idea of that kind, and advises me never to think I have "got there" until the anchor strikes bottom – and here we are, further off than we were then – some further south and a good deal further west. It is agonizing, but it does no good to weep, and anyway, I am not of the weeping kind – but oh! my letters that must be waiting for me there at the consul's![63] I cannot help but mourn. This is the first time we have really been driven back, and so near port, it is worse than it would have been before.[64] Dorothy and I have been on deck for some fresh air. There is plenty of it – we did not need to do as one of the men advised us – go up to furl the royals.[65] He did not advise us, by the way, but the mate – he said there was lots of it up there. Dorothy found a bottle of sewing machine oil that has been set on the floor so it wouldn't fall off the table, and when I found her she *was* a mess. I had just dressed her all up clean and I had it to do all over again. I washed her mouth out with soap and water, but she rather liked it. I don't think it (the oil) hurt her any. She must take her talent for mischief from her father, for *I* never did such things – I was always well behaved.

I read my effusion over the other night to see what we did six months ago, and it rather struck me that you would think Dorothy had been sick most of the time – cutting teeth all the time and cross a great deal. The teeth *have* been on the tapis[66] a good deal, but she has not been cross. Indeed, there couldn't be a better baby, I am sure – and when I look *back* at it, I can see that she hasn't been sick but twice, and then nothing that would have worried me any ashore, but of course I have had to worry a

---

[63] Mail was shipped between ports on small clipper ships that were built for speed rather than for hauling freight, which was how Georgia would have letters waiting for her when she arrived in Japan.
[64] Georgia's desire to get to Japan was amplified knowing that Jack couldn't get rid of the mutineers until they reached the port. She undoubtedly knew that Jack planned to discharge the alleged mutineers in Japan, and they undoubtedly knew they could be arrested and charged with mutiny.
[65] Royals are the topmost sails, which need to be taken in—or "furled"—in high winds.
[66] "On the tapis" is an expression that means "under consideration."

good deal whenever she wasn't just A-1, because I had all the responsibility. There was once I would have given all I possessed for a doctor, but a few baths brought her out all right – she had retention of urine.

She has begun to get fat now, and I hope it will go on, though I must get to work and knit her some more shirts, as hers are too small now. At present she is engaged in pulling papers out of the rack behind me, and coming around to kiss me whenever she tears one.

Oh! for a fair wind!! I shall never again hope to be anywhere in any given time; when I am *there* I will begin to think about getting there – not before.

March 2, Wednesday

I am still of the same opinion – we are still beating around here. We have made a little northing, but it doesn't amount to much, and I can hear the rain pouring, pouring all the time. Yesterday was sunny – the first time in nearly two weeks, but today was so gray as ever, and the outlook for tomorrow is not inviting. I have been finishing my ironing today that I left Saturday because it was so rough – it was a little rougher, but I was afraid they wouldn't keep, so I hung on to my eyelids and toes and made out to get through. I have been trying to knit some mittens for Dorothy, but my needles are too big for the wool, and I am not getting on very rapidly at present – in fact, I have just raveled out for about the fourth time, but I shall keep on. Jocko Verboose de Maginty fra Diavolo has a new house.[67] The carpenter made it for him and it is very pretty. He has a regular little bunk to sleep in – just like anyone. He slips his belt every once in a while, and I find him wandering around. He is only as big as a rat, and would be hard to find, but he *will* talk. I gave him a bath this morning; he doesn't like it a bit, but he does look so cunning and fuzzy after it. I think I'll go to bed, at this rate I shall have time enough to do all the work I want to before I get in.

March 4, Friday

It is six months today since we left Philadelphia, and tonight we sighted the coast of Japan. It was only for a minute, but it was there, and I am feeling quite jolly, though we may be heaven knows how long getting in yet, for a head wind would keep us fighting here indefinitely.

I long so to see you all – my letters will be precious, but I am awfully

---
[67] The monkey.

afraid they will make me homesick. Ah well, we can't have our pie and eat it too. It won't be long before I am home again, and if I had been there all the time I should never know how awfully much I loved you.

March 6, Sunday

We are out of sight of land again, fighting our way, and the weather is so thick you can't see a mile from the ship. The wind set in ahead, and altogether it is horrid. Jack has had about two hours sleep in the last forty-eight, and no sign of any more for a while. I am afraid he will be sick, but he laughs at me. It is raining now, and I am in the deepest depths. Oh! for clear weather and a good wind.

March 13 [Sunday]

Sunday again, and in another hour we shall be at anchor in Hiogo – we ran in and anchored about ten miles from shore last night in a gale of wind and squalls of rain that were something awful. For a week we have been unable to get in at all, and my poor Jack is pretty well used, for there has been a gale going most of the time, and you could never tell one minute where it would come from next, but I hope now that our troubles of *that kind* are over for a while.

About noon a bumboat came alongside.[68] The owner is a Jap and so were the men, but there was an English or American butcher aboard, and he has been giving us all the news he can think of. He says the grippe has been very bad,[69] and I am anxious to hear from you all. He is going to get our letters tonight for us if he can. As I came down into the cabin a few minutes ago I saw the bumboat man looking in. As soon as he saw me, he came inside smiling and bowing and gave me his card. It was fun for me, so I waited. He then produced a number of documents from other captains, and I told him he had better show them to the captain, but he had deeper schemes than that. From inside his robe (he might have stepped right out of the "Mikado," by the way)[70] he produced a pretty little box containing six silk handkerchiefs which he said were for me, *and* would *I* speak to the captain when we got inside – there would be

---

[68] A bumboat is a small boat that carries supplies to ships anchored offshore.
[69] Grippe is an old-fashioned term for influenza.
[70] The Mikado is an 1885 comic opera written by Gilbert and Sullivan. It uses a Japanese setting and characters to satirize British policies and institutions of that period. It saw huge success in both England and America. The Boston Comic Opera Company performed The Mikado in Sycamore in May 1887. Georgia also would have had many opportunities to see the opera during her East Coast travels with her mother.

many more but he was far the best and he would give us 15% off – so I said I would speak to the Captain. I knew he would do just as he pleased anyway, and the handkerchiefs are quite pretty.

8 o'clock – Jack and Dorothy are in bed, and I shall go soon for so much excitement is hard work, you know. When we got pretty well in shore, and after we anchored, there was a swarm of sampans around us, and Japs and Chinamen of all ages and sizes *and* trades descended upon us in an avalanche.[71] Jack was too busy to pay any attention to them, so they turned to me and I have a bunch of business cards that I shall bring home. One, a shoemaker's, gravely informs us "Boots and shoes made. Renewed and the shoe boots". Then the brother of my handkerchief gave us a box of five dozen oranges – they are little, but the sweetest and best I ever ate. They have a rather different flavor from ours, and not a seed in them. The customs officials came off, brilliant in blue and brass buttons, and overhauled and set their seals everywhere. (Just here there was a slight interruption – a belated Jap, a tailor who hadn't got on board in the afternoon, came with his card and a *box of cigars* – I took the cigars and told him to come tomorrow).

Then there was a doctor came off, wants to doctor the ship at twenty-five dollars a month. He was a young Englishman, and a sight for sore eyes – he was so *beautifully* dressed – a little bunch of violets in his buttonhole and altogether the most Christian-looking object I have seen since I left home. But alas! – our letters we could not have – the consul is a mean old thing, and he wouldn't get them, but the butcher promised to get them for us in the morning.

March 14 [Monday]

It is ten o'clock, and that man has not come with the letters yet. I shall have a fit. Jack has gone ashore, and I have been working like two to get things straight. We have had such bad weather lately that we are not in very good order. I have another half-dozen silk handkerchiefs, and two bottles of Florida water, but I am afraid they won't keep it up as they have commenced.[72] Last night about six o'clock the forward cabin door opened and a man big enough to fill the whole place came in with a jolly "hello, folks!" but then he saw me and stopped and remarked "Why, I don't know you!" It was Captain Allyn of the *Titan*, Boston, that lies a cable's length

---

[71] A sampan is a small, flat-bottomed Chinese boat.
[72] Florida Water was a unisex cologne popular in the nineteenth and early twentieth centuries.

from us, and he and his wife (who is with him) have met Osc and Delia somewhere. He didn't stop, but they are coming again soon. He looked very nice, and I shall be glad to see them.[73]

One o'clock – it is *snowing* hard, and it is as cold as we have it at home in the winter. We can hardly keep the cabin comfortable, and Jack has not come with my letters. We heard today that we have been reported lost for a long time. I hope you haven't heard it, but if Jack finds it has been so reported he will cable.[74]

Tuesday morning [March 15]

My letters came about half-past four – eighteen of them – and I read and read and read, and when Jack got off he had nearly as many more, but some of his were business.

They are taking out the oil, and there is a great chattering and jabbering on deck. Just before breakfast a Jap appeared at the door with a bunch of flowers about as big as himself. He is a shoemaker, you know, and he is now making Jack a pair of slippers. It pays to bribe me – the bumboat man and I are fast friends and he is going to find me a girl to take care of the baby. I have given him the monkey. It is so cold the little thing suffered, and I couldn't keep him down here, he is so dirty. I am to have a cat as an offset – ours died.[75]

The row commenced in Allas has revived this morning, and Jack has gone for the consul, but there is very little danger here – there wouldn't be

---

[73] Georgia made the acquaintance of Captain Clarence Howard Allyn (1848–1932) and his wife, Sylvia Delano (Crowell) Allyn (1850–1923). Osc was Jack's brother, Oscar Samuel Yates, who was also a sea captain and co-owner of the *Willie Reed*. Delia was his wife, Delia O. (Hoffses) Yates. She accompanied Oscar on his voyages, just as Jack's first wife did and Georgia was doing now. Oscar had captained the *Willie Reed* before, so it was possible that Captain Allyn was surprised to find Georgia instead of Delia on board.

[74] Reports of the *Willie Reed* being lost at sea were published in several newspapers throughout the United States in the first week of April 1892, after Georgia wrote this entry. The reports did not include specific details, other than to say that the ship was reported wrecked near Kobe, Japan. It is unknown if anyone in Georgia's family read these reports, but an article in the Sycamore *True Republican* dated March 26, 1892, acknowledged that Georgia's family had received word of her safe arrival, the first news they had heard of her since her departure (the article erroneously stated that they arrived in Kioto, Japan, instead of Kobe):

Mrs. Townsend received word this week of the safe arrival at Kioto, Japan, of Capt. and Mrs. Yates. They left Philadelphia in September last on the Captain's three-masted ship, with a cargo of petroleum. They had been over six months on the voyage without touching land, and this being the first news from them, it was naturally received with much pleasure by their Sycamore relatives.

However, it is still possible that Georgia's relatives could have received the false news of the *Willie Reed*'s demise and assumed it happened sometime after she sent the first round of letters.

[75] Georgia never said when or how the cat died. The only other time she mentioned the animal, in the October 27 entry, she called it "the old cat," so it may have died from natural causes. We learn later that the new cat was named Topaz.

that, but the second mate is drunk and not to be depended on.[76]

I did not cable you as the ship was cabled the day we arrived, and I presume Osc would have sent you word before I could have reached you. I hope you have not worried much. This is a splendid ship – she answers her rudder so easily and does so well in bad weather.

I was glad to hear of the new boy, but sorry that Mary had such a bad time, only the baby will make it up to her over and over again.[77] I shall have so many new relatives when I get home that it will take all my time to get acquainted with them. No one said *when* Jen was to be sick,[78] and the paper about Will Denman was not there – how was he killed?[79] Oh! I want to see you all so much, and talk to you, and show you my baby – she thinks Japs are fine and wants *all* the pretty things she sees. She has had a Jap dollie given her, but she treats him as she does her rubber one, and so I have him hung up. The mail goes this afternoon, and I want to write a few words in my other letters, but I will write often while we are here.

Your loving daughter,
Georgia Townsend Yates

---

[76] Georgia was referencing the mutiny in the Alas Strait in January. According to Jack's account from the *Dallas Morning News*: "We reached Kobe without any further outbreak, and I then summoned the United States Counsel Suettius. He investigated the case, and put all of the mutineers in jail. When the ship was unloaded I discharged the crew, and once more breathed easier." Charles Schneider told a different story, which ran in the *Alexandria Gazette*: "When we got to Kobe we complained to Consul Smithers, who told us he had no jurisdiction at sea and told us to go to work. We begged Captain Yates the next morning to keep Turner from us. The answer we got was an order to appear before Consul Smithers. We went to the Consul expecting him to help us. Instead of that he sent us to jail for 10 days each. Paree got twenty days, five days of which was hard labor." The *Brooklyn Daily Eagle* reported a simple version more in line with Jack's: "After arriving at Kobe, the sailors demanded that the captain should send for the United States Consul Smithers. He did so, and after a hearing of the case, the sailors were sentenced to ten days imprisonment and Paree, twenty days." For reasons unknown, Jack did not discharge second mate Turner and kept him for the next leg of the journey.

[77] Georgia was referring again to her nephew, Charles Boynton Townsend.

[78] Georgia's sister, Jennie Townsend Webster.

[79] The identity of Will Denman is unknown, but most likely he was a relative through Georgia's paternal grandmother, Ann Denman Townsend.

Chapter 2: Japan
March 19, 1892–April 9, 1892

In the mid-nineteenth century, the United States strong-armed Japan into signing the Treaty of Amity and Commerce, also known as the Harris Treaty, named after Townsend Harris, the United States envoy who negotiated the deal. The treaty opened trade in three Japanese port cities: Yokohama, Nagasaki, and Kobe, the latter being the destination of the *Willie Reed*.

The treaty gave the United States control over import and export tariffs and extended protections to American merchants, seamen, and settlers in Japanese port cities, who operated under the authority of American consuls. From the Japanese point of view, it gave the United States unequal control of tariffs and allowed American merchants to operate outside of Japanese jurisdiction. This and subsequent treaties came to be known as the "unequal treaties." The Japanese viewed them so unfavorably that the treaties were an instigating factor in the Boshin War, a Japanese civil war fought from 1868 to 1869. The conflict saw the overthrow of the reigning Tokugawa shogunate and ushered in a new era of political and social reform and industrialization, all of which unified and modernized the previously isolated island nation.

Japan did away with the "unequal treaties" in 1899, but they were still in effect at the time the *Willie Reed* arrived in 1892.

Hiogo, March 19, 1892 [Saturday][80]
My dear mamma:

If I don't get at my letter to you, there will be a big gap in it, but since I have been in a week tomorrow I have been busy one way or another until I have done very little writing.

I did not get ashore until yesterday, and then I went with Mrs. Allyn of the *Titan*, and had my first experience of sampans and rickshaws, and I did enjoy it hugely. I have been again today, and I shall keep it up all the pleasant days.

I have a Jap ayah for Dorothy – a woman about thirty who is used to ships as well as babies and she does very well indeed.[81]

I have not made any very heavy purchases yet – you have to take time and bargain about it, but I have seen pretty things enough to make one crazy. And besides that I have fallen in love with the girls – they are so

---

[80] The *Willie Reed* was anchored at the port city of Kobe, Japan, the largest city in the Hyogo Prefecture.
[81] Georgia incorrectly used the term "ayah," which is a nurse maid from India. In later letters, she switched to the correct term "amah," which is a female servant from Eastern Asia. It is also possible that this was an error in the transcription, though this reference appears as "ayah" in both the Sycamore and Seattle Letters.

107

pretty and cute. Mrs. Allyn and I went up to a waterfall this afternoon. On the way there we went through a temple and fed the sacred white pony with peanuts. She seemed to me much like a pet pony in the park at home.[82] The fall was very pretty and everything was new and strange, and the people are so tasty and so pleasant. Everybody bows and keeps on bowing. We went to the tailor's (the ship's tailor) to look at some silk. He is a Chinaman, and his brother's baby was there – the cutest thing you ever saw, and it liked to be tossed and talked to as well as a Yankee baby – it was as old as Dorothy and only about half as big, without a tooth in its head. By the way, my lady Dorothy walks alone and she thinks it is fine, and runs from her father to me and back, laughing fit to kill.

We have our own sampan now, but we shan't have a rickshaw, only get one when we want it.

March 20 – Sunday

It has been so cold today we have not stirred off the ship, and hardly out of the cabin. Jack and I were to have gone on shore and had a long ramble by ourselves, but we didn't want to freeze.

March 22 – Tuesday

Yesterday I was on shore again, but alone, for Mrs. Allyn is sick. She and Captain Allyn are both prisoners, in fact, with bad colds that threaten grippe, but I think they are both better this morning. I went to call yesterday on a Mrs. Lucas, the wife of our consignee. We have been invited to tiffin there twice, but have not been able to go either time.[83] We are invited again for tomorrow, and I hope the third time will not fail. If you people could only see me as I go about here, I would give a good deal. A rickshaw is like a two-wheeled baby carriage with a man to draw it, and they are fine. They are as convenient as a horse and carriage that will drive itself, know where you want to go, carry a watch and interpret for you – all for ten cents an hour. You get awfully cheated here if you pay more than half what they ask you, and I have great fun bargaining for things. They say, "So muchee." I say, "So muchee too muchee – I give *so* muchee," and then they are horrified. I am firm, or perhaps come up a little, but they can't so then I go off and get into the rickshaw, and then they say usually, "Can have." It is great fun. I will

---
[82] Georgia was probably referring to the Sycamore Driving Park, which was built in 1874 to host horse races and showcase the horses of some of Sycamore's wealthier citizens. It was located south of town, on the west side of Somonauk Street, the street Georgia's family moved to in 1887.
[83] A tiffin is a light midday lunch.

show you my cash account when I come home, and it will make you laugh. Jack is coming back to dinner and this afternoon we are going to see the waterfall and take tea at the tea house, and then I shall look for tea sets, as we may go to Nagasaki any day almost now, and this is the best place for china.

We shall know tomorrow whether we will have to go or not – I hope not, this coast is so bad, but I suppose we shall have to go to Hong Kong to load anyway – there seems to be nothing more here.

March 25 – Friday

I have been going about every day and enjoying life greatly. Mrs. Allyn is still sick, but though I miss her very much, I manage to get around pretty well by myself. All the curios they tell you are very old – you know – and Captain Allyn and Jack make fun of us the whole time, but we are hardened.

I have been looking at tea sets lately, and they are so lovely and so cheap. I intend to pay about $15 or $20 for mine, but for Mrs. Reed I can pay as high as fifty, and indulge my extravagance all I like.[84] A short stop just here to entertain the sailors' chaplain, but he was in a hurry and didn't stay long, but I bear up under it remarkably. Last night I stopped on board the *Titan* to see how Mrs. Allyn was, and when Jack stopped to take me up on his way home, they induced us to stay to supper. We sat talking until a little after eight and didn't think much about the weather, but when we started to go we found it was blowing a gale, and we never reached home until after one o'clock. Dorothy was all right, however – the amah had only put her nightdress on wrong side before. I am going to take her over to see Mrs. Allyn and her kitties. It is bright sunshine and will do her good.

March 28 – Monday

The mail goes tomorrow and I must finish my letters. Yesterday we expected Captain and Mrs. Allyn over to dinner but there were snow squalls and it was so cold that Mrs. Allyn didn't dare come out. Captain Taylor came in this afternoon to tea and stayed to supper, and in the evening we all went on board the *Titan* and stayed until ten o'clock.

Today I have been cleaning around a little – ran over to see Mrs. Allyn and have generally been on the go, but this letter-writing must be attended to. I can't make much of writing about the things I see, but I will tell you all about them when I get home.

---

[84] The Seattle Letters have "Mr. Reed" instead of "Mrs. Reed." Either way, Georgia was most likely referring to the ship.

Dorothy sits here by me in her chair while the amah washes out her clothes. She ate a piece of dried fruit the steward gave her and it made her sick all day yesterday. Today she seems all right again except her bowels are a little loose.

It is very cold here yet and Captain Taylor was telling us yesterday that they couldn't get a man to work on the shipyard because thirty-three years ago the prime minister signed the treaty with the Americans and when he came out from the palace the crowd mobbed and beheaded him. That day it snowed, and there had never been a snowstorm so late in the year before. It has never been known again until yesterday, and they all thought something awful was going to happen.[85]

Captain Taylor is very pleasant – he is on here from New York to see about repairing a vessel of which his nephew is Captain, and which had to be sunk to put out a fire that started after the cargo was all in, and she was ready to sail. He is a New Brunswick man, but he has lived in New York for years. There is another New Brunswick man here in the *Kambira* – Captain Brownell, and he is one of the awfully positive men.[86] He was so excited in an argument one day that he became positive that ships were best built of all spruce, which is a rather poor wood – and that is the way he always is – you oppose him a little and he will swear to anything. He is not any too clean, by the way, and his staying powers have never been equaled.[87]

Dorothy says this is no fun – her amah tends her all the time, and she thinks I ought to, as well, but she is pretty well-behaved yet.

When I was over to see Mrs. Allyn there was a curio man there, but he had the worst rot of trash I have seen since I have been here, and his prices were all over the place. I didn't buy anything, but I bought two pairs of vases on board here Saturday – fifteen dollars for the four, and they are lovely. I bought a tea set a while ago of a man who came aboard. It was only half a set – six cups and saucers and six plates, a teapot, cream and sugar, and one bowl, but when I bought it first I only thought I was buying the cups, creamer and not the plates, but when I found he was cheating me I told him he must bring me the rest or I would never buy

---

[85] Captain Taylor was referring to the Harris Treaty (see page 107). The prime minister referred to was Ii Naosuke, a top government official of the Tokugawa shogunate who played a large role in passing the Harris Treaty. On March 24, 1860, he was assassinated by samurai outside the gate of Edo Castle. Even though his signing of the treaty turned many people against him, there were also several complex, internal political factors that ultimately led to his assassination.
[86] John J. Brownell was captain of the British ship *Kambira*.
[87] Georgia's assessment of Captain Brownell proved to be accurate. In June 1894, the British Consul at Nagasaki sentenced him to four months in jail for shooting at the *Kambira*'s boatswain while at sea.

anything from him, and he said he would. He may go back on me yet, however.

Later. I have been on shore all the afternoon, and have about decided on the tea set to buy, but Jack must go and see them first. Then I have several pairs of vases under consideration, and I bought a lot of dolls. One is a beauty and I shall keep it for Dorothy till she grows up.

I think it is colder tonight than it has been at all. I don't know what we are coming to.

I should like awfully well to drop in on you all tonight and tell you the fun I have been having running about on shore, but it won't be long before I shall be there. I am awfully anxious to know when and where we are to load, but that cannot be known yet. We are almost discharged now – a day or two at most should finish us. The ship is way up out of the water and we have to shin up the gangway, it is so steep. They put in some ballast tomorrow.

I want to take Dorothy ashore tomorrow and have her picture taken, but if the weather is as bad as it has been I shan't dare take her out.

Captain Allyn is worse tonight, so he is in bed. I am afraid he has caught more cold. Mrs. Allyn is better, but she is not well enough to take care of him. I must go at my other letters, so goodbye to you all and lots of love and kisses.

Lovingly,
Georgia Townsend Yates

Hiogo, April 3, 1892 [Sunday]

Dear Mother:

Of all frightful holes I ever heard of this is the worst – we have not had two days since we came that there has not been a gale of wind sometime during the twenty-four hours, and today it has been frightful. They were trying to put ballast aboard and one of the lighters nearly sunk alongside. They had to throw all the ballast out and Jack had gone ashore right after breakfast, but it was long after dinner before he could get off. Then the stevedore brought him off in his launch,[88] and when he tried to come up to the gangway the launch hit it and smashed it into kindling wood, nearly throwing two men who were on it into the water – but they didn't stay there long – and altogether it has been something all day, for I have not written half.

---

[88] A stevedore is a dockworker who helps to load and unload the ship's cargo.

I have done very little sightseeing or shopping or anything else lately. Captain Allyn did not take care of himself, and he has pneumonia as a result. I have been there a dozen times a day and all night every night but one this week, so I have very little time to myself, for Dorothy has a certain amount of attention whatever happens. She hasn't waked up a single night when I was gone, so I can leave her with an easy mind.

I had her picture taken the other day and they are first-rate, I am to have them Tuesday, and I will send you one, but the rest of them must wait until I come home. I have all but bought my tea set. I am only trying to decide whether to pay $16.50 or $50 – you can probably tell which way I lean. It is nearly time for another mail, and I am so anxious to hear from you all again – almost as much so as I was when I first came in.

We do not know yet where we shall load, but things rather point to Singapore, and if we go there we shall be as hot as we are cold now, but I don't think anything *can* be worse than this place.

We were intending to go ashore sightseeing today, but the weather put a decided stop to that. Jack and I have just been having afternoon tea all by ourselves, with cunning little Japanese cakes. I wish I could send you some, they are so good. Love to *all*.

Your loving daughter,
Georgia Townsend Yates

P.S. I can hardly wait to hear from Nan and Jen – to think of all the babies, and I can't see 'em.[89]

Kobe, Japan, April 6, 1892 [Wednesday]
My dear Mother:

We are about ready for sea again, and expect to sail tomorrow, so today Jack has about five times more than he can possibly do, and I have quite as much as I want.

I ordered some medicines this morning, and I must go ashore for them about six or half-past, but they are within a block of the steps, and beyond that I am done. I have been packing my things away in one place and another so that they won't fly about.

We shall go from here to Singapore and we have fifty lay days after

---

[89] Nan and Jen were Georgia's sisters Anna and Jennie. When Georgia departed, Anna was pregnant with her fifth child and Jennie pregnant with her third.

we are ready to load, so we won't be home right away.[90] It will probably take us from thirty-five days on to get down there. I am sending Dorothy's pictures by this mail. I send one to each grandmother, the rest must wait until I bring them. I think they are good, only they don't wiggle.

I was glad to hear from Mary and Fred. I will answer it if I can get time, but it is doubtful. I am so sorry I am not there to help bring Charles Boynton up right. I fear they are spoiling him – Mary says they walk with him! I really thought my noble example would have prevented *that*. I had hoped to hear that Anna was well through before I left here. As fast as the babies come have them photographed and send the pictures to me – I want to get acquainted.

My lady Dorothy walks alone all about now. I suppose she will have to give it up when we get out again. She will miss her amah, for I shan't have time to tend her as Toby does.

I have not done a bit of mending since we came in and my stocking bag is something to look at. I can work as hard as I want to all the way to Singapore.

Captain Allyn is very much better. He will be sitting up again in a day or two. I still stay there every night, though there is very little nursing to do. He is very weak and needs nourishment often – that is about all, but Mrs. Allyn likes to have me there for company, and she will be very much alone when I go.

We have all kept well except for very slight colds, and I have some of my lost thirty pounds back. We all send lots of love to you all – especially to our dear Mother.

Your loving daughter,
Georgia Townsend Yates

Kii Channel, April 9, 1892 [Saturday]
Dear Mamma:

I wrote you just before we sailed, and this will go by the same mail, but you won't mind that, I guess. We sailed about 2 o'clock the morning of the eighth but we have had calms or headwinds, so we are not outside yet, and the pilot is still aboard, though he leaves soon.

The weather is fine, so warm that we have the doors and windows open again, and I don't suppose we shall shut them on account of the cold for many a long day.

---

[90] Lay days are the days a ship must stay in port to load cargo. They are determined by the charter company, which advertises how long the ship will be available to load goods. In Singapore, the *Willie Reed* took on cargo for the Brown Brothers & Company of New York.

I hated to leave Mrs. Allyn awfully, and she hated to have me go. I hope she will be at home yet when we get there, as they expect to leave the *Titan* and you people may meet her, if she comes out to the Fair.[91]

All of you write at once when you get this and keep it up until you are told to stop, as we shall be there a long time and I shall long for letters. I have not heard of course from you since you knew we were in, but our letters will be sent on and won't lose much time.

The picture taken of Dorothy in Japan has not been located, but this picture was taken soon after the family returned to Sycamore when Dorothy was around two. (Joiner History Room)

---

[91] Georgia was referring to the 1893 World's Fair in Chicago, known as the Columbian Exposition. Because Sycamore was only an hour from Chicago by train, it became a popular place for World's Fair visitors to stay to avoid the crowded city.

I had to stop for dinner, and since then I have been on deck with Dorothy until now. She is sweetly asleep in her little bed. We have seven of the old men in the crew now and one darky and eight Japs (new) and a cook and cabin boy – said boy is a Jap about 25.[92] They call them all boys there, and though I didn't think much of the idea at first, I think he is going to do very well. He is very willing to be told, and he knows how to do things very well himself.

I was wildly extravagant and bought the fifty dollar tea set, but it is lovely beyond anything. I have lots of pretty things besides, and I mean to get more in Singapore, but what I think as much of as the pretty things are the odd ones

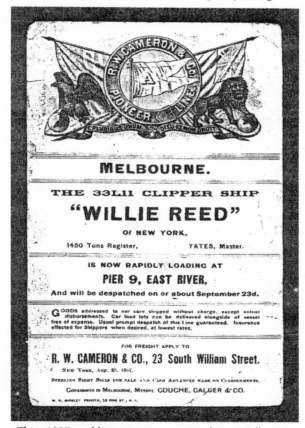

This 1887 public notice was posted in Melbourne, Australia, to advertise that the *Willie Reed* was in port and loading cargo for its return to New York. Similar notices would have been posted in Singapore when the ship arrived there in 1892. (Wendy Jones Smoke Collection)

---

[92] We learn in a later letter that the boy's name was Tarky.

– the chop sticks and the candles. I have a coat such as the coolies wear[93] – the stevedore gave it to me, and a blue handkerchief such as they wear on their heads. Just the day before we left I bought a sword – a lovely one in a little bit of a dark shop. The old woman took a liking to me and brought out the nice things, wrapped in yellow cotton – and I enjoyed it immensely.

    Good-bye, dear people – write me lots of letters and I will "bring you something." I don't know but it would be well to offer a reward among the youngsters. Nothing would make any difference with yours, Mammy – they always come.

Your loving daughter,
Georgia Townsend Yates

---

[93] "Coolie" is a term for an unskilled laborer, usually Indian or Asian. Today it is considered offensive in many cultures, especially if used by a foreigner.

Chapter 3: Singapore
April 15, 1892–August 6, 1892

A busy dock in Singapore, ca. 1890. (author's collection)

In the China Sea April 15, 1892 [Friday][94]
Dear Mother:

I did not intend to keep a letter this time, as I am afraid you folks will get tired of reading it, it's so monotonous, but I find I am lonesome without it, and you needn't read it if you're tired.

We are not quite in the China Sea yet but that's where we will be most of the time.

We have been out a week today, and we have made very good time, take it all around, though we were two days getting out of the channel, and yesterday all day was calm. Today it is rainy but a good wind.

The cabin boy so far is a success. We pay him ten dollars a month, his money, which is only about seven of ours. He does about ten times the work the other man did, though they paid him thirty-five. He doesn't do any cooking, but he keeps the cabin as neat as a pin, does the washing, and takes lots of care of the baby – all this in the pleasantest manner possible. He has to be looked after, of course, but I might have looked after the other fellow day and night and he wouldn't have done half.

I have been mending and am still at it. I had left it so long there is lots to do – then I have been packing away our heavy clothes, some of them not to come out again until we go into Boston or New York.

Dorothy walks all about in spite of the motion. She doesn't like to creep any more. She is good about amusing herself, but she *will* tear any book or paper she gets hold of, and I have to look out for her. She loves to help her papa with the charts, and sits down in the middle of them like a big, extra island.

Our fresh meat is gone so we have returned to salt fare again. I am on short commons, for I don't like it and have to be slowly starved to it.[95]

April 20 [Wednesday]

Yesterday the cook caught a dolphin over five feet long, and we had him for supper and again for breakfast. He was fine. We have been going along very well, and are halfway to Singapore, but the wind has come around a head this morning, and the long half is ahead of us yet.

The weather is getting pretty warm again. The thermometer says above ninety. We were in 18 degrees 2 minutes N. and 118 degrees 32 minutes E. yesterday at four o'clock.

---

[94] The Seattle Letters listed this date as April 16 (a Saturday), while the Sycamore Letters listed it as April 15 (a Friday). April 15 is most likely the correct date because the ship departed Japan on Friday, April 8, and in the letter Georgia noted that "We have been out a week today."

[95] To be "on short commons" is to have only a small amount of food.

April 25 – Monday

For the last three days we have had a good strong breeze, only it is rather ahead. We have made good headway, however, and this afternoon at four we were in 13 degrees 2 minutes N., 115 degrees 27 minutes E. If this wind will only hold we may get somewhere sometime, but life is awfully uncertain.

Dorothy is getting to be so cunning, we just have to sit and laugh at her most of the time. She walks all about, even when things roll. She has learned what candy is, and it is wonderful how well she likes it. She gets very little, however. I am sewing *very* mildly. I don't like to work, and Tarky spoils me. Every morning before we are up everything is swept and dusted and my washing and ironing are off my hands, so I am getting so lazy you won't be able to live with me much longer.

There are five days more in April, and we have been gone nearly eight months. It seems a long time to think of, but it hasn't been so long going.

May 19 [Thursday]

It is pretty nearly a month since I wrote, but topics are scarce. For three weeks we sailed with a fine wind, and at the end of that time we were within three hundred miles of Singapore. It will be six weeks tomorrow since we left Kobe, and we are still two hundred miles from where we should be. Dead calms, with now and then a breath of air, but what we make one day we usually lose the next, so it is slow work. We have had a very light wind aft for the last twelve hours, so I hope we have made a little something. We are 107 degrees E., 2 degrees 50 minutes N., so if you look us up you'll see there are islands all about us, and that makes it dangerous. The water is alive with fish, but they won't bite, and this morning I saw my first sea serpent. He was very young, I think, for he was only three feet long or so, but he was undoubtedly a sea serpent, and will probably grow to immense length when he gets older.

Dorothy is a regular little fish for water – she has two salt water baths a day with a fresh water one thrown in occasionally to clean her, and you should hear her yell when I take her out. Only let her see a little puddle on deck and she is in it before you can wink. She is altogether the worst mischief I ever saw.

The mosquitoes nearly eat us up – I don't know what they'll be when we get in.

May 22 [Sunday][96]

We are slowly nearing our destination, or rather were nearing it by jerks. One day we make a little, and then we have two or three days dead calm. Today is one of the last.

Dorothy has been sick a little once or twice since we have been aboard here, but never enough to make her uncomfortable at all, and otherwise we have not known a sick hour, until last Friday when I took cramps in my stomach, and they made themselves very unpleasant. I am better this morning, so I can sit up again, but I don't want any more. Jack and Tarky have taken care of Dorothy. I don't know what we should have done with the old steward, but Tarky is so good.[97]

It is bright and sunny today, but awfully hot because there is no wind. We were 106 degrees 3 minutes – 1 degree 50 minutes yesterday.

Note: The following letter was from Jack to his mother-in-law, sent "in haste" the day after the *Willie Reed* arrived in Singapore. He sent it while retrieving Georgia's letters from her family. She remained on the ship due to her "illness." The Sycamore Letters and Seattle Letters included it after Georgia's May 29 entry, but I have moved it here for the sake of narrative chronology.

Singapore, May 24, 1892 [Tuesday]
Dear Mother:

We arrived here last night – 45 days from Hiogo. Georgia is on board, and I don't know as she will get her letters ashore for this mail, so I drop you a line that you may know where we are.

I have but a few minutes to mail time, and Georgia has the letters on board, so I don't know any news from you, but hope all is well.

I wish you could see the Baby. She runs all over and we of course think she is the one of all.

My regard to all, in haste,
Your Son,
John E. Yates

Note: Georgia wrote the following letter separately from her collected entries (April 15 through May 29). She claimed that those letters were "not

---
[96] The transcriber of the Sycamore Letters made a note that this entry was written in pencil.
[97] This illness was the first indication that Georgia was pregnant.

ready" and she wanted to send something upon her arrival at Singapore, which is why some of the information is repetitive. Eleanor would have received this letter before the previous six entries and the four entries that follow. I have placed it here for the sake of narrative chronology.

Singapore, May 24, 1892[98]
Dear Mamma:

We are in port again at last, but I have been sick for the last week and my letters are not ready, but there is a mail going tonight, and I want to send a line.

The first three weeks out we had lovely winds, and came on to within three hundred miles of here, and since then we have had nothing but calms and very light head winds.

Later. They brought the letters off from the consul's, but there were none since you knew we were coming here – only those forwarded from Japan. There was news enough in them, however – *two* more boys! I shall not be able to realize my blessings if you keep on. Do send me some pictures of them – surely Charles is old enough to be photographed. I received a photo of Frederick from Jen, but it was broken.[99] He looks as if he had grown a good deal since I saw him.

Dorothy has found her language at last – she speaks Malay like a native, and will no doubt do all our talking for us. We have a lovely basket of fruit – oranges, mangoes, bananas and pineapples – brought off as a bribe. I shall be as adept at taking bribes as a ward politician by the time I get home.[100] I am not equal to much writing this morning, for I have been going about a good deal, and I am as weak as French soup. Jack had to bathe and dress the baby while I was sick, and he did very well, but it came hard on him to have so many things on hand at once.

We are having a heavy rain squall – they tell me it rains here most of the time. The weather is not to be depended on. I can begin to look forward now to the time when I shall see you all again. We will be here probably between two and three months, and it will take four or five to

---
[98] This letter was undated in the Sycamore Letters.
[99] The two new boys were Georgia's nephews Edward Denman Claycomb (son of Anna Townsend Claycomb), born March 20, 1892, and Pierce Webster (son of Jennie Townsend Webster), born March 27, 1892. Charles was her previously-mentioned nephew Charles Boynton Townsend. Frederick was another nephew, Frederick Charles Webster (son of Jennie Townsend Webster), born December 13, 1888.
[100] Georgia might have been taking a dig at her brother, Fred, who at that time was a ward alderman in Sycamore, the first of many elected positions in his political career.

get home, and that is a very long time to give us. I fully hope to see you all before Christmas. And the dear babies – how can I wait? Give my love to all, and I will write soon.

Your loving daughter,
Georgia Townsend Yates

May 25 [Wednesday]

Day before yesterday we dropped anchor here about three miles off shore, and here we are still. I am about right again, only pretty weak. My letters came off quite early, and I read with joy of my two new nephews – wish I could see them, and oh! *somebody* send me their photographs.

It rained most of the time yesterday, but today is bright, though there is no telling how long it will stay so. Last night about six a big fleet of fishermen came out and we watched them setting their nets. Their boats are only two or two and a half feet wide, and are twenty feet long, but the fun is to see them in the morning when they drag the nets in toward the shore, for they beat the side of the boat all the time to drive the fish in. There must have been a hundred of them, so it sounded like a carpenter shop. The boats going about are of all kinds, the one that looks oddest to me is what they call the bamboo sail, which is this shape [see image] with bamboos running up the seams to hold it out.

Tomorrow we go in to have the ballast taken out, then we go into the

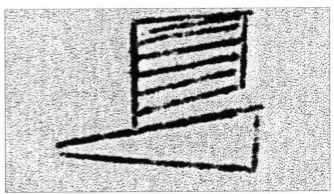

The transcriber of the Sycamore Letters included this small sketch in the May 25 letter. We can assume it was copied from Georgia's original drawing. (Joiner History Room)

dry dock to copper and then for cargo.[101] I don't know how long it will take.

The flags are all flying ashore today, as they are celebrating the Queen's birthday. I am too far off to see much, but I expect Jack will tell me about how they look when he comes back. The *Lucy A. Nichols* left here this morning for New York.[102]

May 26 [Thursday]

This morning we towed in to take out our ballast. It rained hard just as we got in and has only stopped, but they are about to begin now. We have an awning now, so I have been on deck in spite of the rain, getting a good deal of amusement from the shore. The general scantiness of attire is rather a shock at first, but I soon got over that — in say fifteen minutes at most, but when a gentlemanly Chinaman filled a tub with water, and removing his trousers (his only garment) began to take a bath, I thought it was going rather far, but even then I could have stood it, but he proceeded to take off his *hat*, and I looked no more. Most of the coolies are dressed for *extremely* hot weather. Jack has gone ashore and won't be back until night, I suppose.

May 27 [Friday][103]

Last night Captain Wilson, the wharf manager, took us to call on his wife. They live only a block from here and have been married two weeks. She is a pleasant little English girl and came out to be married.

I have invested a dollar in corals and shells — I have enough to fill a bushel basket nearly, and yet I have been badly cheated. They bring great boatloads alongside, and they are lovely.

Our washerman is a Malay — an old man with white hair, all dressed in white, with a great white turban, and yet the name of this patriarchal-looking person is Billy Buttons. I enclose his card, and you can get some idea of his importance. He charges four cents, however, and that is a cent more than I paid in Japan.

It is going to rain again. In fact, it rains most of the time, I guess, and

---

[101] To "copper" a ship means to fasten copper plates to its hull to prevent corrosion. At the time, it was an expensive but cost-effective process that reduced repairs and insurance costs.

[102] The Lucy A. Nickels was built in Brewer, Maine, in 1875. In both transcriptions, the ship was spelled "Nichols" instead of "Nickels," the latter being the correct spelling. Georgia may have chosen the former spelling because the ship's captain was C. M. Nichols, who was not related to the ship's owner, J. C. Nickels.

[103] The Sycamore Letters attached this entry to the previous entry, May 26. The Seattle Letters have it as a separate entry dated May 27, which is how I have separated it here.

I am lazy or restless. I want to go ashore, but I can't take the baby very well when I have no idea where I am going, and I have no nurse yet. I expect the doctor will be aboard today, and I shall ask him about it, I guess. I put in my time writing letters, only I have none to answer.

Singapore, May 29, 1892 [Sunday]
My dearest Mamma:

Ever since your letter reached me – day before yesterday – I have been trying to write, but every time I think of Anna, and I have to put down my pen. There is nothing to say, and I can only cry and cry. I try to think it is only a little while till we'll all see her again, but it does no good. I want her here and now. And the dear babies that she loved so much – no one can ever make up to them however much we try for what they have lost.[104]

Your letter is dated the day we left Japan, and I never had the least feeling of what you felt – it seems as though I ought to have known. Of course I was anxious, but after I had your first letter saying that she was doing well I never worried any more, and I could not believe it. I wish I could be there, but I hope to be within six or seven months from now.

Dear baby knows I feel badly, and though she is so busy with so many things to see, she comes every few minutes to give me a kiss and a hug – she is very well and takes all my time to look after her.

We are in the dry dock now and will be most of this week. They have offered us a house to live in on shore while we are here, and I think we shall go, but take our meals aboard here.

It is raining, or rather pouring, and has been all the morning. I am afraid it will keep on all day from the way it looks.

Write me about what you are going to do. Poor Frank[105] – I will write him if I can. How is the dear little baby? I wish I could be with you to do something, but if I can help in any way when I come you know I will do anything I can.

I sent my letter written before I heard it – sounds heartless, but you understand, I did not know.[106]

Your loving daughter,
Georgia Townsend Yates

---

[104] Georgia's sister Anna Townsend Claycomb died on Friday, April 8, 1892, at the age of twenty-eight. She suffered from complications after the birth of her fifth child, Edward Claycomb, on March 20. One can only hope that Eleanor did not see or believe the early April reports about the *Willie Reed* being shipwrecked near Kobe. If she did, she may have believed that she had lost two daughters in the span of only a few days.

[105] Anna's husband, Francis "Frank" Claycomb.

[106] She was referring to the letter dated May 24, 1892.

THE LETTERS

Singapore, June 6, 1892 [Monday]
My dear Mamma:

Another week has gone by and we are all coppered and today they have sent us the first of our cargo. Our lay days are up the 24th of July, but we hope to get away a little before that.

It was a long hard week for Jack while we were in the dock – he had to watch everything, or they would half do it, and then those Chinamen steal everything they can lay hands on. The first day we were in the dock Jack fired the second mate [John Turner]. He was drunk *all* the time we were in Japan, and he commenced again here. I was glad to see him go, for I had always disliked him.

The sailors were drunk most of the time, and now we have only four of the old crew left, and three Dutchmen we shipped here. Sunday morning we left the dock at five o'clock, and the cook hadn't come aboard, and what's more he didn't come until this morning, and there we were, for the boy had had a vacation and is sick with the fever, so you can see that I had my hands just about full. The crew cooked for themselves except the bread, which I worked and put in the pans, but I cooked for the cabin. I am glad to say the new cook came in time to get dinner.

I have not been ashore yet except to walk about the dock a little, and I don't think I shall see much of Singapore at this rate. I am trying to write on a lap board on deck, but Dorothy is knocking it about so I doubt you will be able to read this.

I had a nice packet of letters Monday, but all by way of Japan. We haven't had any direct yet except from Oscar and mother.[107] I am so sorry Frank got that letter as he did – it must have hurt him so, and he has had so much to bear. I think so much of the babies and wonder how you have arranged things. Lily is a dear good girl to help you.[108] No one could do it so well. Give her my love and tell her I shall bring her something pretty, because she is good to you.

Angie's finger was probably all well before I heard about it, but I am just as sorry,[109] and my dear mamma, you must be careful and not overdo. No one can do things as well as you, but let them do the best

---
[107] Jack's mother, Sophia Yates, and brother, Oscar Yates.
[108] Lily was most likely Lillian Walker (1870–1921), a friend and classmate of Georgia's cousins, Daniel and Elinor Wild. Several articles in the Sycamore *True Republican* showed that she regularly socialized and sometimes vacationed with the Wild and Townsend families.
[109] Angie was Angeline Allen (ca. 1827–1902), a long-time house worker for the Townsends. She began working for Eleanor in the early 1860s, before Georgia was born.

they can, for you must not be sick.

Give my love to all the dear people, and tell them I will try to write, but that I am very, very lazy.

Your loving daughter,
Georgia Townsend Yates

P.S. I had a long letter from Mary. What a nice letter she writes.

Singapore, June 12, 1892 [Sunday]
My dear Mamma:

The deck is far the pleasantest place this afternoon, and with a young person as lively as Dorothy about, ink is dangerous, so please excuse pencil. She has taken the hot weather as usual to cut some teeth. She has two upper double ones through, though only three lower ones yet. She doesn't seem to mind it much, but she is rather cross for her.

This morning the dubash brought off an amah for me.[110] She went off again to get her clothes, but I am expecting her back any time now. She is a Chinawoman, and I like her looks very much. Dorothy went to her at once, but she will go to any Chinaman or Malay that comes along. A white man she won't speak to.

I have been ashore once, but I found it too hard work with baby. Now I shall try to go more.

Yesterday we had a very bad squall. We just got the awning in in time. We dragged a little. The *Daniel I. Tenney*, the only other American ship here, lost her awning.[111] A native boat capsized and one man drowned – quite near – though we didn't know about it until afterwards.

There have been two German men-of-war here all week. They went this morning – it made something going on. That and the shell and coral boats were my only excitement. I have quite a few shells and some pieces of coral, but mostly small because they will be such a bother to carry.

We have had a good deal of trouble about a cook, but I think we have a good one now, and as Tarky is getting well I hope things will go smoother.

I think of you all at home so often, and wish I could see you, but it

---

[110] A dubash is an interpreter.

[111] The *Daniel I. Tenney* was captained by George S. Wilson. The ship was named for Daniel Ingalls Tenney (1800–1881), a native of Newburyport, Massachusetts, who became a jeweler in New York City and catered to the city's elite. He later donated much of his fortune to charitable causes and his hometown. The ship was built in Newburyport in 1875.

won't be very long now, I hope. We are not getting cargo very fast, but there is a German boat loading by the same firm for New York, and we will get more when she is gone.

Captain Wilson of the *Tenney* is very pleasant. He is a man over sixty, and has been going to sea most of the time since he was thirteen. His wife is not with him, but he has three dogs, two cats, a parrot and a monkey in the cabin for company. He came out to Madras and has some lovely things he bought there.

Dorothy has just put a score of kisses in the corner, and she is adding a great many hugs. She is so sweet and she keeps us busy. Jack sends lots of love to you all, and I can never get enough home letters.

Your loving daughter,
Georgia Townsend Yates
x x x x x x x x
x x x x x x
x x x x x
x x x x
(kisses)[112]

Singapore, June 15, 1892 [Wednesday][113]

My dear Jennie:[114]

I have started ever so many letters to you, but I could never get them written. I cannot write about Anna – I cannot bear to think that she will not be there with the rest of you when I get back.

And so you have another boy, and I was not there to come down and make believe take care of you. You will have your hands full, but Marion will soon be going to school – think of it, when it was only the other day I went down to take care of her.[115] And Dorothy is almost a year and a half old – such a great girl as she has grown to be. She has just cut her first double teeth, and she is so full of mischief it takes one person to look after her.

It is very warm here, though we have about as comfortable a place as there is, here on the water. I have only been ashore once – it was too much like work to take the baby, but yesterday the dubash brought off an amah for me, and now I shall get about more.

---

[112] The transcriber of the Sycamore Letters typed these x's, so we can assume that Dorothy's original "kisses" were marked in a similar pattern.
[113] The Seattle Letters list this date as June 13.
[114] Georgia wrote this letter directly to her sister Jennie Townsend Webster.
[115] Marion was Jennie's daughter, Marion Webster, born January 21, 1887.

We have very little cargo as yet, but we hope to get away before our lay days are out, and then for home. It seems a long time since I saw you all – it is over a year since I left Sycamore. It will be hard to go back with Anna gone.

It is a great care for mother to have a little baby to care for. I hope Lily will stay with her – she is so helpful. If Frank can get a good woman to take care of the house it will be a great help, but such a one as he needs is hard to find. I don't suppose I should be much help if I were there, but I wish I could do something. Dear Nannie – how shall we do without her?

Mary wrote me such a nice long letter. I will try and write to her this mail. I hope she helps with the children. She could give you your liberty a good many times. Have you the same girl or someone better? I have

Georgia's sister Anna Claycomb passed away on April 8, 1892, while Georgia was still at sea. (Joiner History Room)

wished a good many times that I had brought my Japanese amah with me – she would have been glad to come and if I had not needed her someone there would have been glad to have her.

We hope to have home letters today, as we think the French mail came in yesterday. I haven't had any direct yet, only those forwarded from Japan. Jack has had one from his brother.

Later. The mail came out. I only had one letter, so I shall only send what I have written and not write any more. Jack says he is going to write to Charlie, but I doubt if he does – he isn't good at letter writing.[116]

Your loving sister,
Georgia Townsend Yates

Note: The following is another letter from Jack to his mother-in-law.

Singapore, E.I. June 20, 1892 [Monday][117]
Dear Mother:

Georgia had an invitation to luncheon and to ride today, and has gone on shore. And as she did not have any letter written and the mail closes at 6 P.M., I am a poor substitute – but if I don't write you a few lines you won't get any letter by this mail. I wrote you a few lines when we first arrived, then we had not heard of Anna's death. I know that on such a subject words don't amount to much, but still I know that when great sorrow comes upon us it seems good to us to feel that we have those who sympathize with us, and I can only say that you and Frank and all to whom she was dear have my sympathy.

Georgia and Dorothy are both well. We had a nurse for a while, but she left yesterday. I don't think much of the nurses here but Georgia can't leave to go on shore without one, and we don't like to take her out too much in this hot weather. Summer and winter are much the same here as this place is almost upon the Equator. We have hardly begun to load yet and it will probably be the middle of July and perhaps later before we get away from here. One other American ship here, and I expect she will get away some time before us, and then we shall be left alone to fly the flag of our country.[118] A ship bound from Madras to San Francisco called here a few days ago, and Georgia saw the first American woman she has seen

---

[116] Charlie was most likely Jennie's husband, Charles Webster.
[117] The E. I. stands for East India. At the time the *Willie Reed* visited Singapore, the port city was under the administration of the British East India Company.
[118] He was referring to the *Daniel I. Tenney.*

for some time. She had a letter yesterday from her, sent back by the pilot, and a little present in it for Dorothy. I can tell you, when you get so far from home, Americans draw together. We liked Japan very much better than here, and if we could only have loaded there we could have seen the country and had a much pleasanter time, but the state of business is such that there was no chance to choose. I expect if Dorothy should walk in at your door today you would hardly know her. She goes up out of the cabin and on top of the house herself. I expect everything is looking beautiful around Sycamore now, and even the old dry hills of Maine are clothed in green. No country my eyes ever rested on is so beautiful as our own.

Remember us with love to all the dear friends in Sycamore and Galesburg.

Your son,
John E. Yates

Singapore, June 26, 1892 [Sunday]
Dear Mother:

Yesterday I had such a nice long letter from you – the second that has come directly here. We have been here over a month now and have only two or three light loads of cargo in yet. I am afraid they will keep us all our lay days and they are not up till a month from today. The *Tenney* is nearly loaded – and we shall be so sorry to see her go, for she is the only other American ship here, and Captain Wilson is so pleasant.

My amah only stayed a week and I was not very sorry to part with her. She was careless and took baby into the sun, and baby didn't like her very well. I have only been ashore once this week, and then it was to spend the day with a lady ashore. She is the wife of a hotel keeper here. They are German, and he is an old captain. I had a pleasant time, while Dorothy was delighted, as there were two little girls of three or four.

There are three gunboats, two French and one English, lying close to us, but they are not unusual so they are only a bother.

We have had a good deal of fever aboard. The men don't take any care of themselves, and sleep out in the dew and all that. Jack has had two light attacks, but baby and I have not had any. Baby is cutting teeth yet. Two of her upper and lower double teeth are through, so it takes most of my time to entertain her. I received the pictures of Fred and Charles. He [Charles] is a fine baby and looks as if he might weigh about as much as Dorothy does now. I should think he would be a handful for his mother. Dorothy can say a number of words – she says her name is Attie (Lassie) and she can't make anything of Dorothy.

I think Pierce will be a very good name for Jennie's boy. She can't very well name him Daniel as he has Webster to go after it.[119] I will keep that for my second boy.[120]

I am glad you have a good woman at Frank's, and hope you will keep her, and I am so sorry about Angie's hand.

In writing about the new houses, you say nothing of Fred's, so I suppose it must be given up for some reason, and you will have them with you. We do not know what we shall do, but I think it is very doubtful if we live in Sycamore – probably in Maine for a time at least. You will not need us, and Mother Yates has the next claim. I suppose Jack would say the first, and though we will not live with her, we will be near.

We are going to New York – there is no doubt about that now, and our cargo, if we ever get it, will be gambier, cutch, rattans, sago and tapioca and other things of like nature.[121]

I do not like it here, we lie so far from shore it takes two hours to get off when the wind is ahead, but such little things have to be put up with.

I am in hopes to get a tiger skin tomorrow, or at least Jack is going to see if they have come, and then he will stay home and take care of Dorothy while I go to see them. I want more than one if I can get good ones. I have one good leopard skin, and I am trying to buy another, but they want too much for it and I can't get ashore to argue with them. Jack is no good at that sort of thing.

June 28 [Tuesday]

Yesterday we had a long letter from Mother Yates, and one from Oscar, and a bundle of papers from you, and today I am getting a big lot of letters ready to go. I write more than I get, but that is not your fault, you dear mamma – as I am always sure of one from you.

I haven't any amah, and I want to go ashore. I think I'll have to leave Jack with Dorothy, only he doesn't seem to relish the idea. We had cargo yesterday and some today. I only hope they will keep it up.

---

[119] Pierce Webster was named for his great-grandfather, Daniel Pierce. Georgia didn't want him to share a name with Daniel Webster (1782–1852), the famous politician known for his fiery oratorical skills and support of strong federal government.
[120] This line implied that she already had a name picked out for a first boy. When she and Jack eventually did have a son, they named him John, after his father. They did not name any of their sons Daniel.
[121] Georgia was listing oils and spices derived from tropical plants native to this region of the world.

I have been here as long as I care to be, and I should like to get started home – it's an awful while since I have seen you.

Your loving daughter,
Georgia Townsend Yates

Dorothy and her cousin Charles Boynton Townsend. This picture was taken in early 1893, soon after the family returned to Sycamore. (Joiner History Room)

When Georgia departed, her brother, Fred, was planning to construct an elegant new home in Sycamore. The project got underway in May 1892. Fred spared no expense on the Queen-Ann-style home, which was built by local contractor W. H. McAlpine, who later built DeKalb County's courthouse and the Northern Illinois State Normal School (present-day Altgeld Hall of Northern Illinois University). The house is pictured here, ca. 1900. Today it is on the National Register of Historic Places and operates as a bed and breakfast. (Joiner History Room)

THE LETTERS

Singapore, June 28, 1892 [Tuesday]
My dear Mary[122]:

I have meant to answer your letter long ago, but if you were here you would understand that a person does very little she doesn't *have* to do, and things get put off.

I had an amah for Dorothy, but she wasn't very good, and only stayed a week. It is so hot I don't dare take Dorothy out, so I have not been ashore in over a week now, and there is no knowing when I can go.

I am afraid that crepe can't be had. I didn't get it in Japan – there were so many things I didn't do that I wanted to, but if I had known what a poor place this was to get things I should have tried harder.

I have had a necklace made of silver and tiger claws for Dorothy, but as she won't want to wear it for some years to come, I presume you will get some good of it. It is very pretty besides being so odd. It is rather tiresome here – I can go about so little, and I am afraid we shan't get away before our lay days are up. They are putting in cargo today.

Mary and Fred sent me some pictures of Charles. He looks like a fine boy, and I am awfully anxious to see him and the other babies. If only Anna could show me hers instead of the poor little one with no mother.

You must be at home again by this time, and Dan must be through with college altogether.[123] Jack and I have been wondering what has become of Fred's house. No one has said anything about it, so we suppose it must have gone up some way. There are lots of things we want to know that we don't find out, but we shall be home before long and then we shall ask questions until you'll all wish we would go to sea again.

I expect we shall settle down to live in Round Pond and then you must come and make us a good visit.

Your loving sister,
Georgia Townsend Yates

Singapore, July 5, 1892 [Tuesday]
My dear Mamma:

Yesterday was the glorious Fourth, and we had the ship dressed with every inch of bunting we possessed and fired off a lot of firecrackers and such-like stuff, but nevertheless it didn't seem in the least like the Fourth of July.

---

[122] This letter was addressed to Georgia's little sister, Mary Corey Townsend.
[123] Mary would have been away at Lombard College in Galesburg, Illinois. Dan was their first cousin, Daniel Pierce Wild, who graduated from Lombard while Georgia was at sea. He returned to Sycamore and took a job as a vice president at his grandfather's bank, working under Georgia's brother, Fred.

We had three or four letters from home last week – one that came by San Francisco from you. There is a mail due today, and we hope for more, for we never have enough.

You write that Fred's house is under way and that is the first word that has been said about it since we came away, so it is no wonder we thought they had given it up. You told of all the new houses but not a word of that, and I haven't heard from Fred or Mary in ages.

Yesterday afternoon we went ashore and went to their museum. It is very good. They have a particularly fine collection of butterflies.[124] I have an amah again, an Indian woman this time, and a grandmother, though she doesn't look thirty-five. She is a vast improvement on the other and I can trust her very well. She has a pleasant way of taking care of me as well as the baby that will quite spoil me. She brushed my hair ever so long last night, till I was so sleepy I had to go to bed.

Dorothy grows sweeter every day. I wish you could have seen her yesterday. There had been some men here and I had bought a couple of small crepe shawls – one was pink and black with colored embroidery, and I put the dark one over her shoulders and she had a little silk handkerchief they had given her in her hand, and she marched up and down with more airs and graces than a peacock, and wiped her nose on her handkerchief until we were nearly dead laughing at her.

We are not getting loaded very fast. I expect we will have to stay our time out and that is three weeks from today. The *Tenney* is almost loaded and will probably go this week. I have been doing some sewing while I have been tied here, but I don't get on very fast, I am too lazy. I shall have to work when we get outside and into cooler weather, for Dorothy needs clothes.

I intend to write a lot of letters today, though I have few to answer. I don't know what I'd do if yours didn't come so regularly.

Your loving daughter,
Georgia Townsend Yates

Singapore, July 11, 1892 [Monday]
Dear Mamma:

Last week I wrote seven letters and expected to get a lot, and there wasn't one for me. This week I had four but I am only going to write this one.

There has been lots of coming and going among the ships here this

---

[124] Georgia and Jack visited the Raffles Library and Museum, which opened in 1849 and moved into a grand European-style building in 1887. It has continuously expanded ever since and today operates as the National Museum of Singapore.

week – the going hasn't affected us as they were Germans and Italians, but among the comers are an American and a Nova Scotian. The Nova Scotian has a lady on board, but though she is very pleasant she is no good for running about, as she has three children already and expects the fourth any day. I don't envy her.[125]

The *Tenney* will go tomorrow. I expect we shall feel badly to see her go. Captain Wilson has been so pleasant, but of course he is glad to start.

But the saddest thing of all I only found out the other day. Our lay days aren't up until the third of August – eight days later than I supposed – because they don't count Sundays, and I know they won't let us off a day before our time.

Dorothy caught a heavy cold – the doctor says she caught it from her father, and she hasn't been like herself for a day or two. She is better this afternoon, and into mischief with most of her old vim.

I am surrounded with bills on all sides, so I hardly dare move an eyelash. Jack is making up the vessel's accounts to send home, and if anything should be moved there is no knowing what might happen.

There was lots of news in the papers this week. I think the last lot from Round Pond must have lost itself. We haven't had any from there in so long.

It looks a long time yet when I count up the months before we can get home, but they will go as the rest have. I haven't bought a very great lot of stuff here as it is high, and I don't like most of the things, but I have a few. I mean to get some chairs this week if I get ashore.

Give my love to all the folks. Jack and Dorothy send theirs as well. Dorothy says she wants to go and see Grandma.

Your loving daughter,
Georgia Townsend Yates

Singapore, July 18, 1892 [Monday]

My dear Mamma:

Another week is nearly gone, and mail day is coming around again, so I must go at my letters.

Yesterday I let the amah go ashore, and I didn't know what to think when she didn't come back at night, but this morning she came and said her sister was dead, so I gave her money and let her go again. She said she would be back tomorrow. I am so sorry for her – she was telling me about

---
[125] This was Mary Ellen (Curie) Douglas (1858–1900), wife of Captain Jonathan "Jack" Douglas (1858–1902). Their ship was the *Calburga*, a Canadian bark built in Nova Scotia in 1890.

her sister only yesterday before she went ashore.

Dorothy is getting pretty well over her cold. She coughs yet in the morning some, but she feels pretty well. Yesterday afternoon we went over to the *Calburga* and she had a fine time with the three children there. When we went on deck this morning we found a German ship that had sailed for New York about a week ago anchored near us. She has sprung a leak and will probably have to discharge her cargo and go in dock. It will be an awful job and cost a big amount of money as well. They are pumping all the time to keep her clear.

We bought a last lot of shells this morning – also four pairs of deer's horns and the saws from two sawfish. I begin to think we had better settle in New York for how we are ever to get all this truck beyond there is beyond me.

Two weeks from Wednesday our lay days are up, but if they don't hurry I don't believe they'll have us loaded then. They won't keep us long on demurrage, however.[126]

Dorothy has been asleep nearly three hours, but it has been pretty quiet and nothing to disturb her. She is such a cunning darling, but I am afraid she won't prove to be very well-behaved when I take her among folks. She has been too much alone.

Jack has gone ashore and won't be back until tonight. I was going, but of course I can't leave Dorothy without Susanna.[127] I haven't been off for a week. Only two more mails to send letters by, and then I come myself. Send me a letter as soon as you receive this to St. Helena.[128] We may touch there as we go very near.

Your loving daughter,
Georgia Townsend Yates

Singapore, July 24, 1892 [Sunday]
My dear Mamma:

After thinking the matter over in every light, I have decided to use Molly as a halfway post office to get my news to you a little before I get in, but so you will only have a little while to worry as I fear you will, though there isn't really the least reason for it.

---

[126] Demurrage occurs when a ship is kept beyond its scheduled lay days. The company that chartered the ship would have to pay the owners for the extra days, which was why Georgia believed they wouldn't be kept long.

[127] Susanna was probably the amah.

[128] St. Helena is a small, remote island in the South Atlantic Ocean, located approximately 1,200 miles west of the African country of Angola. It is infamous for being the island Napoleon Bonaparte was exiled to from 1815 until his death in 1821.

THE LETTERS

As you probably suspect by this time, we expect an addition to the family about the middle of next February[129] and we have decided, after a great deal of thinking, to take some kind of furnished place in New York or Boston and stay there until about next April, when we will come on West to see you all. And I hope, mammy, that you will be able to come on to see us when we get in – that is the reason I write, for I do want to see you so much, and I shall not be able to come out before Spring.

What we shall do after next summer is still undecided – when we thought Fred and Mary had given up their house we decided to live in Maine for a few years at least, but you have always wanted us there, and now you have so much extra care, though I think with two children I might be more bother than help – still, we will decide nothing until we get home and try to see how we can really help you most.

I think, by the way, I would rather you considered this letter privately, and didn't let anyone see it – it strikes me that it would be more modest not to offer until I'm invited, but you have said often enough that you did want us, and then I am never afraid of your misunderstanding me anyway.

As for the small affair of No. 2, I hope, dear, you won't worry.[130] I am quite well. I haven't had as many unpleasant feelings as I had before, and I shall be about through with them before I leave here. We are charmed with the idea ourselves. Dorothy wants a baby, Jack wants a boy, and I think it is time to be getting on with our thirteen.[131]

When you come on, will you bring what clothes I left there? There are only a few of them – I shall make what more I need on the way home. I wish I could see you to talk to, it is so much more comfortable, but it won't be very long now. I have certain hopes for Thanksgiving, but that is doubtful – it all depends, you know. We have had a long, tiresome time here, and shall be very glad to get away. It is an odd thing that every woman that has been in harbor since we have been here has been busy with family cares – the first one had her baby and has gone, there is one expecting every day, and the other one is about the same as myself, so I

---

[129] Georgia seemed to be calculating her pregnancy back to when she had been sick in May. Based on when she gave birth (December 22), and assuming the babies didn't come too early, she probably had been pregnant since late March or early April.

[130] She was referring to baby number two.

[131] An ambitious plan! Such a specific number implies an inside joke, or she was using it due to its characteristic as a superstitiously unlucky number. Georgia didn't quite reach thirteen children, however, but she eventually gave birth to eight children, far more than any of her siblings.

am not alone.

Dorothy and Jack say you are to come as soon as we telegraph. They want to see you *almost* as much as I do.

Your most awfully loving daughter,
Georgia Townsend Yates

Singapore, July 26, 1892 [Tuesday][132]
My dear Mother:

I received a long letter from you yesterday and from your reports all along I should think you were having more than your share of rain.

I am glad you were able to get down to commencement. It isn't often that one of the family graduates, and all that can ought to be there to help when they do.[133] The babies' proofs were pretty well faded, but I could still see that they are fine-looking children. Edward doesn't look very fat beside Charles but he is a bright-looking baby. Charles is a fiery-looking piece, he is growing to be very fine-looking but I don't see a striking resemblance to either parent.

I am sorry you people set your ideal so high in the matter of letters. It takes nearly three times as long for a letter to go from here than from

This is not the photograph that Georgia received on her voyage, but it shows all three of her nephews who were born during her voyage. It was taken when they were one year old. Left to right: Edward Claycomb, Charles Townsend, and Pierce Webster. (Joiner History Room)

---
[132] This is the last entry in the Seattle Letters.
[133] This was a reference to her cousin, Daniel Pierce Wild (see footnote 123 on page 133).

Japan, so you must have had an anxious wait.

Expect us home in time for Christmas, and then I don't think you'll have any waiting, and if we get in earlier than that it will be a mistake on the right side and won't worry you much.

The Paulsens were over last evening and we had such a pleasant chat. They have four children with them and a small cabin at that, but they get on as nicely as possible.

We expect the *Calburga* out again today, she has been in dock over a week discharging coal, and the *Wakefield* in a day or two. She is in dry dock being metaled, so we shall be quite lively. Captain Minot of the *Wakefield* is a Californian and one of the kind who thinks there is no other place like it. He is very pleasant and he doesn't like this country any better than we do.[134]

Dorothy is as fine as ever. I wasted most of yesterday morning putting pockets on one of her new frocks. I am going to get some pink calico to make her some frocks to wear home, as her sea clothes are about done up.

I am afraid I have come to the end of my news – we have eight more lay days – that is working days and the ship is far from loaded. I rather hope they will have to pay some demurrage for keeping us here this length of time.

Captain Minot was born and raised in Maine, so he and Jack would have had stories to share. His wife also traveled on many of his voyages around the world, but on this trip she had stayed in California. This image ran with the account of his 1892 voyage in the April 15, 1893, issue of the *San Francisco Morning Call*. (author's collection)

---

[134] The *Wakefield* was a bark built in 1873 in Newburyport, Massachusetts. It had sailed out of San Francisco and arrived at Singapore by way of Australia. The ship departed Singapore a few weeks after the *Willie Reed* and sailed to New York following the *Willie Reed's* route. An account of Captain Minot's voyage appeared in the April 15, 1893, issue of the *San Francisco Morning Call*. Here is an excerpt describing Captain Minot's time in Singapore:

> Here the *Wakefield* rested for two months while the barnacles were being scraped from her bottom and some new copper added. Then for the home voyage to New York. Oh, home again! How willingly the crew hurried along the loading and getting ready for sea again. They loaded with coffee and spices, sago, gambier and rice, and all the staple things that come from the far distant shores of sunburnt Singapore.

Probably there will be only one more mail for us to receive and send letters, but we'll send from Anjer and perhaps from St. Helena.[135] And oh! I shall be so glad to see you again.

Your loving daughter,
Georgia Townsend Yates

Singapore, August 2, 1892 [Tuesday]
My dear Mamma:

Here it is, the second of August and here we are and as far as I can see this week will see us no further. There is one comfort however, tomorrow is our last lay day and after that we have demurrage. There are two lighter-loads of cassia alongside that Jack refuses to take they are so torn and broken.[136] Probably if they weren't in such a hurry to get us loaded they would have taken them ashore and repacked them, but as it is they sent them right here from the steamer. But the ship is responsible for damage and Jack won't have them.

I haven't been ashore much but I have been visiting a good deal with the two ladies in the harbor – the one I like best goes today or tomorrow to Hong Kong.

I bought Dorothy some tin dishes yesterday and she thinks they are the finest things she ever saw. Anything more breakable wouldn't last her long.

We had four here to tea Friday and six Sunday evening – it quite filled our table up.

Last evening we were over to the *Calburga* a while.

I will write again before we leave and then the next thing I telegraph from New York.

Your loving daughter,
Georgia Townsend Yates

Singapore, August 6, 1892 [Saturday]
My dear mother:

We are about ready now to put to sea again, and we are only too happy to go. They are doing the last bit of loading this morning, and most of the stores are aboard. Jack is about done up – you can't leave

---

[135] Anjer (or Anyer) is a port town on the west coast of the Indonesian island of Java. The *Willie Reed* stopped there as it passed through the Sunda Strait.
[136] Cassia is a type of cinnamon.

a thing to anyone's honesty, as there isn't any in Singapore. Everything must be gone through and weighed – almost everything has been short-weighed, and a lot of poor stuff that had to be picked off and sent back in it – it is hard enough work getting ready for sea anyway, and thinking of everything without everybody trying to cheat you at the same time.

I am getting things put away pretty well and as we won't have any bad weather probably for a few days I hope nothing will get broken, but being in port so long things get about.

I am sorry to leave on Mrs. Douglas' account. She will be so all alone – not a woman except her ayah and it is so hard to be sick so. However, she knew what she was likely to encounter when she left home, and she can be thankful she isn't out at sea with no doctor or anything else. I think I will stay at home for that sort of thing myself. We have the crew aboard and they are a very decent-looking lot of men, most of them are from that Norwegian ship that was lost in Gaspar Straits a few weeks ago.

I expect I shall want to get out and push lots of times, but after you settle down at sea the days slip by pretty fast. Jack thinks we will be about four months going, but of course it may be more or less. We will telegraph as soon as we get in and we shall expect about half the family on at once. Also that there will be plenty of letters waiting for us. Our New York address is c/o Yates & Porterfield, Ship Brokers.

My lady is just waking from her nap, *later*, and puts in a kiss for grandma. She would like to write some, but mamma thinks the letter looks badly enough as it is.

Your loving daughter,
Georgia Townsend Yates

Dear Mamma:[137]

We have been waiting since three this morning for wind enough to leave, but there is not a breath and I fear we must wait until night.

Jack is pretty tired but the worst is through now, and I feel as if we were really started.

The ayah was just over and she says Mrs. Douglas is feeling badly. I hope she will get through quickly.

Your loving daughter,
Georgia

---

[137] This brief note and the one following were included with the previous letter.

10:30: We were happy too soon – it is a dead calm and the current has set us out again. There are two bumboats alongside and at present the entire crew are chasing the chickens that crawl out of the coop as fast as we put them in. Oh for a wind!

1:30 – A three-inch breeze, but it is useless to keep this agony before you, so I will seal this. We are sure to get through sometime.

Ever your loving,
Georgia

Chapter 4: Return Home
August 15, 1892–December 9, 1892

The China Sea [August 15, 1892 – Monday][138]

My dear Mamma:

This is our eighth day out and we are only a very short distance on our way yet. The first four days out it rained continually and we never got the sun once, so we had to go awfully carefully among all these islands and reefs, and since then we have had very light winds, but we knew we should have a hard pull until we get out past Java Head, for it is the S.W. monsoon now.[139] We are all well and Dorothy and I think it is fine fun to take care of each other again. Jack has to be awake about twenty hours out of the twenty-four, but he'll get over that when we get into clearer water. After we pass Anjer we have very little in our way.

I have learned a great deal from the ladies with children I have met, and I take things a good deal easier than I did, though I never worked very hard. I find that it is pure foolishness to iron, and I doubt if I do any at all. Things get stretched and pulled while they are drying and get along the best they can with that. Aren't you shocked?

We haven't had wind enough to blow the mosquitoes away yet, and they are an awful pest.

I am getting some sewing done, but the weather is pretty warm, and I don't hurt myself.

Dorothy says ever so many words now – the only ones she puts together are "Mamma come." She could talk I know if she would, but she makes us understand so well without that she won't take the trouble. She is very fond of the kitten, and as it won't defend itself I can't keep her from abusing it rather.

Tuesday morning[140] – my young lady got up from her nap just there and had to be dressed and fed. Last night we had a dead calm, so we anchored, so as not to lose what we had made, and they are just getting under way again. While breakfast waits and I am nearly starved – how I wish we had Anjer astern.

August 20 [Saturday]

We have been having a hard fight not to go to the northward, but

---

[138] This letter was undated. Assuming the *Willie Reed* departed Singapore on Sunday, August 7 (the day after the previous letter), then Georgia wrote this entry on Monday, August 15, based on her comment that this was their eighth day at sea.
[139] Java Head is a bluff at the southwestern end of the Indonesian island of Java. It serves as a landmark to passing ships that they are either entering or exiting the Sunda Strait. The Sunda Strait is the body of water between the islands of Java and Sumatra that connects the Java Sea to the Indian Ocean. American newspapers erroneously reported the Sunda Strait as the site of the mutiny aboard the *Willie Reed*.
[140] August 16.

we have a very good wind tonight. I only hope it will hold. I have been making myself a new wrapper out of some old brown cashmere I had with me, and it is really quite pretty. I have it all done but some eyelet holes, and I should have had it finished long before noon if my young lady had not been possessed of the very imp of mischief, and so kept me after her from the time she woke up this morning until about fifteen minutes ago when she went bye-bye, thoroughly tired out. She has learned to go over the forward doorsills – about eighteen inches high – faster than I can go after her, so we have to keep the rough weather boards in – nearly as high again – and I expect she will be over that by tomorrow. She runs off with spools, scissors, and continually takes her father's dividers to stab the cat, and carries that unoffending animal around by its head, its tail, a handful of skin or anywhere she happens to get hold. She tears the magazines, takes my chalk that I am marking buttonholes with and marks all over the settees,[141] and as a windup I found her in the bathroom with the cover off her chair, calmly washing her face in the potty. After that she was tied up in a chair and so carefully watched that she did little more mischief except bathe the cat in the tub of water on deck, and steal a bottle of pepper sauce out of the pantry window.

She has been looking like a little prizefighter for two or three days because she got up in bed in the middle of the night and the ship rolled and threw her down. In the morning her eye was as black as you please and today it is just getting to the interesting green and yellow stage.

Jack is getting to look like a ghost – he hardly gets any sleep and things won't be any better until we get through these places. We are close in to the coast of Borneo tonight and we hope for a land wind that will give us some southing before morning. We have a very good steward and things are a good deal better cooked than they were coming out. You are having warm weather yourselves by this time – nearly the end of August and Dorothy is nearly nineteen months old. This time last year we were in Philadelphia, and I was buying a lot of stuff most of which I have never wanted at all, as was the case with about half the other stuff I bought. Another time I'll know better.

August 25 [Thursday]

We are feeling pretty well for us, this morning. Last night we got a start that took us through a narrow passage we had been trying to get the <u>better</u> of for three days, and this morning we have a little wee fair wind

---
[141] A settee is a long seat with a back that sits two or more. It is basically a small couch or love seat.

– our hopes fly high. The weather is all kinds – one day rain enough to swamp us and the next not a cloud. Yesterday was hot – no wind. We were anchored all day until five o'clock. We are just getting under way now, and the men at the windlass are insisting that "We bound for Rio Grande".[142] Dorothy likes to hear them sing – she has her other lower tooth through that has hung back so long, and I think the second set of double teeth are starting through. The eye teeth don't show yet.

August 29 [Monday]

Still struggling, but we still hope to some small extent that we shall be home before Christmas, but you needn't worry if we don't. If we go on like this we'll be lucky to get home by spring. Yesterday we had fresh fish – not a very big one, but very sweet and boney.

This morning Topaz caught a rat – that is nothing new, for he is a splendid little ratter, though he is hardly larger than a good-sized rat himself, but Jack and the steward caught him and before they could get the rat away from him they had to choke him. I think it was cruel – you should have heard him howl.

Today is fine and I am going to wash a lot of things that I want the starch out of to pack away – Jack's white coats and so forth. I hate to build a fire but it can't be helped.

August 31 [Wednesday]

Yesterday Dorothy was nineteen months old and a greater rascal of that age can't be found. We have just had a grand excitement – caught a kingfish about three and a half feet long – fresh fish for supper – don't you wish you were here?

We have had a very good wind for the last few days, but very light – we are in 106 degrees 44 minutes E. and 2 degrees 58 minutes S. I have just worked the four o'clock sight, and we will soon be in sight of the light on North Watcher – it is a great deal more comfortable when you have a light to steer by among a lot of islands.[143]

Dorothy's tooter has been hung away in the sloproom about ever since

---

[142] Georgia was referencing the sea shanty "Rio Grande," also known as "Bound for the Rio Grande" and "Away, Rio!" For lyrics, see Appendix B.

[143] North Watcher is the northernmost island of Indonesia's Thousand Islands, a chain of 110 islands northeast of the Sunda Strait. The light Georgia was referring to came from a 157-foot cast-iron lighthouse built on the island by the Dutch in 1869. Today, it is the oldest Dutch lighthouse in Indonesia still in operation.

she learned to walk, so when I took it down today it was as good as new, and she enjoyed it hugely for half an hour, but it is hung away now as she soon tired of it. She can get about to suit herself better with nothing to aid her little legs, or rather check them. She wears that black shade hat you gave me and she won't go on deck without it. I wish you could see her with her little pink calico frock, her little bare feet and this black hat knocked into a dozen angles, strolling about to see the "tuchas" (chickens) or Mrs. Dennis the pig. When her father looks at the chart she is right there and lies flat on her little stomach and travels her wee fingers about in the wisest manner possible. Jack scolds her for a bother, but he has to laugh and kiss her, she's so sweet.

September 1 [Thursday]

We are past North Watcher, but it has been a thick rainy night with the wind ahead, and it doesn't show much signs of changing yet. Dorothy is feeding the cat by cramming pieces of toast into her mouth and otherwise making herself agreeable. When I ask her what she is going to do when she sees grandma, she clasps herself tight with both arms, and gives a tremendous grunt and then goes to kissing at the greatest possible rate – so you must be prepared.

September 2 [Friday]

We are quite near St. Nicholas Point, but the wind is dead ahead and it's doubtful if we get to Anjer today.[144] All day yesterday it rained and poured with heavy wind squalls, in one of which about a third of our foretop gallant sails blew clean out of the bolt ropes and is probably somewhere on Sumatra now. At night it cleared and we hoped for better weather, but at one o'clock it shut down again and is going yet.

8:00 o'clock – I have just come down from on deck and the water is covered with pumice stone. Mr. House says he never saw it so before and he's been through no end of times. I hope they haven't had another earthquake.[145]

---

[144] St. Nicholas Point is on the northwest point of Java. When the ship passed St. Nicholas Point, Georgia was in the Sunda Strait.

[145] Indonesia is located in the highly active Pacific portion of Earth's Ring of Fire. An example of the region's volatility occurred in early June 1892, while the *Willie Reed* was anchored at Singapore: Mount Awu, a volcano on Indonesia's Sangihe island chain, erupted and killed over 1,500 people. What Georgia and first mate Eldridge House most likely witnessed were the materials released by nearby volcanic activity. They were not far from Krakatoa, a volcanic island in the Sunda Strait that erupted in 1883, killing over 36,000 people. The newspaper account of Captain Minot's voyage (see footnote 134 on page 139) included his trip past the obliterated island. Captain Minot sailed through the Sunda Strait just a few weeks after the *Willie Reed*, so he and Georgia would have witnessed similar scenes. Here is the newspaper rendering of the captain's report:

September 3 [Saturday]

We are getting in [to Anjer] this morning if it does not rain – two things to raise our spirits. And now it is goodbye again until we reach New York sometime before Christmas, I suppose. I have no hope for Thanksgiving. Be sure you are all ready to come when I telegraph, for I shall be in a state of mind to see you, and bring along any of the rest that can come. Jack and Dorothy send dearest love.

Your ever loving daughter,
Georgia Townsend Yates

The Indian Ocean, September 6, 1892 [Tuesday][146]
My dear Jen:

I hope that mamma will be on to see me before a letter would have time to reach her, so I will send this to you, as I am so used to having a letter about that I should hardly feel natural without one. If anyone ever had a hard and generally provoking time we did in getting from Singapore to and through Sunda Straits – it took us an even four weeks, and when you understand that with a fair wind it can be done in four or five days, you can see it was no fun. After we were clear of there however, and it really seems some time ago already, although it was really only yesterday morning about ten o'clock – we took the wind – we have sailed along finely between yesterday at four P.M. and today that time we made two hundred and fifty miles and that is a big day's work.

When I worked the sight today at four we were 100 degrees 56 minutes E. and 8 degrees 10 minutes S., heading S.W. by W 1/2 W with a fair wind, fine weather and all well. It makes New York look weeks nearer.

I have been darning stockings today all day. I have only done what I had to for ever so long, and my bag was full, but it is empty again now. I am having an awful time with my tiger skin – it is more trouble than a pair of twins[147] – when they kill the beast they only stretch the skin and don't poison it, and as the mate has used up all the arsenic, I can't fix it now.

Coming through Sunda Straits they passed the island of Krakatoa, where the big earthquake and volcanic eruption occurred a few years ago. They found the fires from the bowels of the earth still flaming and smoking, making a grand and awful sight in the blackness of night and a clouded horizon in daylight. They found half of the island cut right off sharply and sunken out of sight entirely. Once it was a fair green isle, rich in tropical vegetation. Now it is as barren and bleak as a hillock in Death Valley. *San Francisco Morning Call*, 15 April 1893.

[146] The *Willie Reed* made a brief stop at the port city of Anjer (or Anyer), where Georgia mailed her last letter. The ship departed again on Sunday, September 4.

[147] You may call this line a fluke or foresight, but Georgia was, in fact, pregnant with twins.

First the worms began on it around the ears and among the hair – I used spirits of camphor and have got rid of them, though there is no telling for how long – and now I find the ants have commenced on the inside. Jack laughs at me, but I will save it yet.

There is a great sea running, and I have had to hustle a bit to get things solid. It is very different out here from in the China Sea. Pierce must be about six months old – he will be a great boy before I see him.

September 8 [Thursday]

Still flying along in fine style. Just now we are making twelve knots. Yesterday morning I had a fried flying fish for breakfast. He flew aboard, and he was excellent, but small. I wished there was more of him. We have four bunches of bananas and we shall kill ourselves if we keep them from spoiling. I think some of them will go to the chickens. Yesterday I nearly finished a set of underwaists for Dorothy. I don't think they are a good shape, but I had to cut them from a small one and I was never much good at that. We are having glorious sunsets and moonlight nights. They were never equaled – I wish you could be here for a few days to enjoy it – I don't suppose you would want to stay longer.

September 14 [Wednesday]

We still travel on in good shape. Today was the smallest day's work since we left Sunday, and that two hundred miles. The weather begins to be colder and Dorothy has stopped going barefoot and looked up her flannel clothes. I have been very industrious, though most of it had to be in the knitting line, and there has been a big sea running and things have been standing on their heads a good deal. Dorothy didn't like it the first day but she has her sea legs in fine working order now and she doesn't give a – --- – for the roll.[148]

I have knitted a new set of shirts for Dorothy and two or three to be for No. 2 and today I have just finished a pair of mittens for Dorothy. They are the first I ever knit and as my needles and my yarn didn't match my rule I have had to make them up a good deal. However, they will keep her warm. I have been hemstitching today – went clear around the skirt of a frock for No. 2 – wasn't that pretty well? I shall finish that frock tomorrow, but there are plenty more to do, and how would you like to sew on the machine when you had to hold on with one hand first to keep from

[148] It is unknown whether Georgia or the transcriber omitted this word.

being pitched over it and the next minute to keep from being fired across the room. I am so used to it now that I really think I shall put my machine on rockers when I get home so as not to miss the motion.

September 18 [Sunday]

Sunday, and really the most tiresome day in the week. Jack objects to my playing solitaire, and I don't work, so it is long. I never did like to read on Sunday anyway – I haven't been to church now in over a year.[149] Two weeks tomorrow since we lost sight of Java Head, and we are a long way from there – haven't taken the sight today, but yesterday at four we were at 22 degrees 15 minutes S by 61 degrees 30 minutes E. This morning – and in fact I believe it's there yet – a sail on our starboard bow, but too far away to tell much about it. Some old bark or other.

Twelve weeks from now ought to see us in, but heaven only knows if it will or not.

September 23 [Friday]

I have just been on deck exercising – a thing I don't do nearly as much as I ought – and I found we have an albatross astern – the first this way. I shall have to begin fishing for them tomorrow, but tonight it was too late. Fine winds still, but not so strong – made 180 miles today. Our position at four o'clock was 44 degrees 32 minutes E. by 28 degrees 41 minutes S., so you can see we are getting on. We have raised and then passed three barks one after another – we should like to pass a ship but none have come within our range so far. I continue to sew but I don't make very rapid headway. My knitting gets on better. Wish we could see you all tonight. I hope you have as fine weather as we are having – it will be cold enough by the time we get there.

October 3 [Monday]

We are still this side of the cape, or rather we are south of it. Our fine winds have all played out the last week and we have enjoyed a series of calms between head gales, and I am afraid [we] would have made a little less than nothing except for the current that runs to the westward here and gives us a little every day. Last night was awful – it blew a heavy gale from the west and that against the current kicked up a sea that would make your hair stand on end. All the loose furniture was lashed fast but it managed to do a good deal of bumping for all that, and any stray book or anything that we thought was

---
[149] This was Georgia's first and only mention of church.

secure ranged from one side of the cabin to the other. Jack put me to bed about half past six for fear I might hurt myself some way (he never shut an eye all night) and there I lay, most of the time on the board put up at the side to keep me in, and heard things fly, but it calmed down some before morning, though we have a rather lively sea here all the time.

Dorothy and I take a run on deck two or three times a day, and her ladyship has got her mittens pretty well black, but she likes them. I manage to get a good deal of amusement out of the crew – watching their peculiarities, and then the mate tells us lots of their yarns. The boatswain is a little stubby chubby stub-nosed Russian Finn and Mr. House tells us that he claims to be of noble birth. He says he is a baron in his own country. He also travels about the country with a magic lantern show, but I believe got drunk and made a failure of that, so he shipped before the mast. He had just been wrecked before he found us, and the yarns he tells of the sailing qualities of that vessel are wonderful. Her captain had been in fourteen shipwrecks and had lost eight vessels himself. The steward (a West Indian darky) is good at his work but I need a dictionary at hand for the big words he uses. The men in general are very fair of their kind. There never was a sailor yet suited with his grub, so ours are on allowance now. The worst one is a darky (the only one) and he has so much to say that he is a nuisance. He had a row with the mate tonight while I was on deck. They fight among themselves sometimes when one wants to boss, and then they play cards for tobacco. One man was boasting on Thursday that he had enough to last him two years, but alas for luck – on Saturday he was buying more. They buy that, and clothes, of the captain.

October 6 [Thursday]

Time, 6:30 P.M., and behold me – having my infant to bed and sat down to write you again. I am feeling well – better than usual because my supper was fine. Very few sat down to just such a supper I am sure – they stewed breast of a mollyhawk and the liver of an albatross – and don't you wish you had some?

I have been picking feathers and down all the afternoon – it is a job, but the down is lovely and the feathers not bad. Jack wants the light, so good-night.

October 16 [Sunday]

We are ten days nearer home than when I wrote you last, and a very good ten days too. The next day we passed the Cape, and took the S.E. trades, which should carry us pretty well up to the Line. Not a strong wind, particularly, but always fair and fine weather.

Day before yesterday we crossed the meridian of Greenwich, and today at noon we were 22 degrees 19 minutes S. Lat. It was cloudy at four o'clock so I don't know just where we are tonight, but we are getting well toward St. Helena. I want Jack to stop but he thinks it would be more bother than it would be worth for so short a time. I cut Dorothy's hair today – quite short, and now everyone calls her Bobby. When you ask her where Bobby is she puts her hand on her head. I have finished four new frocks for her since I wrote – that makes six altogether. The last one is white silk with a lot of lace, and I think myself that it's awfully pretty. I shall make her two or three more, as she is very short. She hasn't had anything new except sea clothes in over a year now. I have just finished knitting a sack for No. 2 – it is white yarn and blue silk and the stitch is very pretty, but I didn't get a very good shape on it – it was a lot of work and I wish it looked better. I have started another, all white, and I hope to have better luck. A little over a week and I shall be twenty-six. I begin to feel old, or at least as if I ought to feel old. Jack gave me a crepe shawl before we left for a birthday present.

October 25 [Tuesday]

Ha! I am growing old – twenty-six today if I figure rightly. I never thought of it until just now, and as I don't see any way to celebrate I will finish my mending. I had fried flying fish for breakfast, anyway.

We are still sailing nobly on – the winds have been a little stronger for the last day or two. Yesterday we made one hundred and seventy miles and at four were in 19 degrees 48 minutes W. long. and 9 degrees 27 minutes S. Lat. The old ship has her head toward home and she smells her oats. The weather is warm and fine and we have a ship in sight – she is lofty and carries skysails, and I am afraid is outsailing us a little. She is too far off to signal.

6:30 P.M. Lat. 7 degrees 49 minutes S. long. 22 degrees 7 minutes W. Dorothy is in bed, though not asleep, for I can hear her saying "Ta-ta-mamma, ta-ta-papa, ta-ta-lucka-lucka (chicken), ta-ta-ow! (cat)," and so on. She talks all the time and puts words together a good deal, but she has so many words of her own manufacture that I am afraid no one will understand her at home. I have had a great disappointment today – I have saved up our last cocoanut for weeks – two at least – for my birthday, but alas! when it was opened it was empty – the only bad one of the lot. I haven't gotten over it yet.

October 30 [Sunday]

We near the line, but the S.E. trades haven't gone back on us yet, though they may at any time.[150] Dorothy is twenty-one months old today. I wish you could see her, and more than that, that she could see the children. Dear baby, she doesn't know what a child is, hardly – I wonder how she will like No. 2. We have struck the course of the outward bound ships – there is one in sight now, and day before yesterday we saw three at once. By the way, the big one I wrote of the other day didn't pass us after all – we passed it, which is a great deal pleasanter – to us, at least.

I have finished Dorothy's frocks – eight in all, and now I must try for a hood. I never have any luck with that sort of thing, but she has nothing to put on her head. I expect the cloak I made before we reached Japan will be awfully short. She has grown so tall, but it will have to do for this winter.

We are painting and polishing and generally getting ready for port at a great rate. I wish you could see us outside, we look so fine, but in the cabin we are only just commenced. Fine weather isn't so necessary here.

Oh well, it is time for baths. I wonder how soon I will see you after we do get in. Don't get Pierce into trousers before his auntie sees him.

November 8 [Tuesday]

Today is the election, and I expect the excitement runs high, but we might as well be calm for we can't know anything about it until we get home.[151] We are having fine winds and weather now. Yesterday at four we were in 45 degrees 19 minutes W. Long. and 12 degrees 33 minutes W. Lat., having made two hundred and fifty miles in twenty-four hours. Yesterday was our third anniversary. We have never been able to celebrate one yet. The first one we were on our way from Boston to Sycamore, and the last two at sea.[152] We have flying fish for breakfast

---

[150] According to an article in the *New York Herald* (published on December 10, 1892), they crossed the equator the next day, October 31.
[151] Georgia was referring to the 1892 presidential election between former president Grover Cleveland and incumbent president Benjamin Harrison. The pair had faced off previously in 1888, when Harrison took the presidency from Cleveland. Cleveland persevered in 1892, making him the first and only US president to serve two non-consecutive terms.
[152] During their first anniversary, Georgia and Jack were traveling to Sycamore so Georgia could give birth to Dorothy at her mother's home.

nearly every morning now – two or three – and we see hundreds of them. Such pretty things, looking as if they were made of silver paper, rising out of the blue water, sometimes one or two and sometimes in schools of a hundred or more. They are like bits of foam, and no one who has not seen it can have any idea how blue the sea is when you are really on deep water. Along the shore it is always more green than blue. I haven't been able to wash yet this week – the ship is jumping about so, and I have been losing strength a good deal the last month. Lack of exercise, I suppose, as I am perfectly well. I wish we could be in for Thanksgiving, but that, though possible, is very far from probable, as these winds are not likely to last always. I can hear my small daughter talking away to herself, though she was put to bed to go to sleep near half an hour ago.

November 13 [Sunday]

Sunday again and still a good wind, though not so strong. Yesterday we were in 61 degrees 12 minutes W. and 21 degrees 27 minutes N. but there is lots of space between us and New York yet with no one knows what kind of wind. It is very warm yet, warmer than is usual they say, but I have no doubt we shall find it cool enough north of the Gulf Stream.

I am trying to give Dorothy some idea of her cousins, but I am afraid my success is small, though she can almost say Marion, can say Nellie pretty well, and calls Charles "Char-char". She has never seen a baby younger than herself since she has been gone, and very few children at all.

November 17 [Thursday]

And though it isn't *calm* it is a great deal nearer to it than we enjoy. I greatly fear that we shall fall short even of *one* hundred miles today.

I have been having a terrible time trying to get a hood made for Dorothy. My struggle would fill a volume, and even now I can only hope that it wouldn't be noticed in a crowd, but I never could do that sort of thing.

It is still very warm so that only yesterday I put a petticoat on Dorothy. She still runs barefoot and until now has worn only a skirt and diaper besides her frock. She doesn't care about clothes.

November 21 [Monday]

Weather much cooler and wind behaving far from well. We have had two days of strong N. W. wind, and though we have not lost we have made almost nothing. Today we are getting on again a little, but we can't go straight where we want to yet. Thursday is Thanksgiving – how I should like to spend it at home. It seems a little aggravating to be "so near and yet so far". We have donned our heavy clothes, and this time for the rest of the winter, I suppose.

Dorothy is such a great girl now, though her face has not changed much since she was a baby. She is two feet eight inches tall and weighs twenty-three and a half pounds in very nearly what her Uncle Fred calls live weight – and she sings a most fascinating little song of her own composing. I want to show her to you all, for we are awfully proud of her – in fact, I feel as Charlie did before Frederick's time – that girls are good enough for me.[153]

I spend a good share of my time now calculating how long it will take mamma to get to us after we get in. Jack says that patience is a virtue I am lacking in.

November 24 [Thursday]

Thanksgiving night, and though I know I have much to be thankful for, I had hoped to have more, and I am afraid I can't be thankful for one of the worst northwesters they often get up – even here – in fact the wind has been northwest or else wanting for a week now, and we have made very little over a hundred miles in that time. Hard luck, isn't it, just when you're almost in. The glass is rising tonight however, and the wind falling somewhat, so I hope again.

Dorothy is in bed and asleep by this time. She sleeps about twelve hours solid and a short nap during the day. Her hair grows so fast – it is only a little while ago I cut it tight, and now it is about ready to cut again. It grows very dark – I think it will be black as her father's was – a good deal darker than mine. Her eyes are dark, too.

Three times lately we have hooked a fish, and every time he has got away – isn't that hard luck? However, with all our trials it is a comfort to think we are within five hundred miles of New York, and if we ever do get a wind it won't take long to get there.

---

[153] Georgia was referring to Jennie's husband, Charles Webster. Jennie and Charles's first child was a girl, Marion (born 1887). Frederick was their second child (born 1888).

December 8 [Thursday]

And here we are, off Sandy Hook,[154] as nearly as we can make out in a mist as thick as to be nearly as bad as a fog – Jack has had two hours' rest in forty hours and over – I don't think there will be much left of him when we get in. It will be three weeks tomorrow since we had a wind at all favorable, and we have gone through about everything in the meantime. Four months ago today we left Singapore.[155]

December 9 [Friday]

We are really here at last, anchored off the Battery in sight of Brooklyn Bridge, and under the uplifted arm of Liberty. To say that the landscape is different from that I have gazed upon for the last four months doesn't entirely express the charge. After I wrote yesterday the fog shut down thick as butter and never cleared until after dark. We got a pilot aboard about six and then the wind came ahead and we were all night beating up to the Hook where we got a tugboat and about half past eleven anchored here. Jack went right ashore in the tug and I must wait until night for my letters. I have been hard at work all morning, and the cabin looks quite gay with my tiger skin and our other rug on the floor and lots of things hung and stuck and scattered about, but I have plenty more to do and a pile of newspapers that I have only glanced through – when I get time for them.

8:00 P.M. Jack has come back – about five o'clock – and as I only had two letters – one from you and one from mother – I have worn them about out reading them, besides reading the Round Pond letters several

---

[154] Sandy Hook is a barrier peninsula that juts out of northern New Jersey.
[155] Captain Minot (see footnotes 134 and 145 on pages 139 and 149) arrived at Sandy Hook several weeks after the *Willie Reed*. Here is the end of his account that ran in the April 15, 1893, issue of the *San Francisco Morning Call*:

> The gallant *Wakefield*... had good winds until the Cape of Good Hope was sighted. Here a long and steady gale kept them from rounding the cape for a good two weeks.... New Year's day they were only 240 miles off Sandy Hook. Forty-eight hours more, they hoped, every man alive on the good bark *Wakefield* would be sipping his grog before a rousing fire in a bowery saloon. What if the weather was biting cold and the ship covered from stem to stern with a coating of ice? What if four of the small crew were below decks suffering with Java fever? What if the sea were running mountains high and the head winds blowing a steady gale, and the icebergs drifting perilously near? Soon the gale would go down and then home, ah sweet home, would be reached. But the gale didn't go down and home was not made—no, not for a whole month. For thirty long, cold, frozen, stormy days, the gale lasted, with intervals of abatement less than twenty-four hours in duration. And as often as the *Wakefield* gained fifty knots headway toward New York just so often she was driven back into the fogs and mists of the gulf stream. It was not till the last day of January that Sandy Hook was finally rounded and a pilot sighted. "What made it so bad for us was that we had been sailing for the past twelve months in tropical waters," says the captain, "and the extreme cold nearly froze the marrow in our bones. Such a climate is only fit for the Esqimaux," he added, contemptuously. "I wouldn't live in such a place."

times and now I must get some letters ready to go in the morning.

We go into dock tomorrow afternoon and the next day I hope that mother will be with us.

I have taken excellent care of myself, and though I am very far from being strong as I was before I am perfectly well – I lay all that to the lack of exercise.

As to coming home, I do not know yet what I shall do. If I were as strong as I was before I would not go, for I think the house there is full enough without my taking a very mischievous baby and the prospects I have into it, but I don't know now how it will be. Jack is too tired and has too much business on hand as yet to talk of what is best.

We were surprised to hear that Mr. and Mrs. Webster were so nearly living in their new house – it will be pleasant for you to have them near you. [156]

From what you write I am afraid I shall hardly know the children – the idea of Marion's owning a thimble. When it is only the other day I went down to take care of you and her – but it is nearly six years ago.

You will never reach the end of this if I don't stop soon, so goodnight and tell Mary that as I didn't have a letter from her I shan't write to her for a day or so. Give my love to the folks and try to imagine how much I want to see you all.

Your loving sister,
Georgia Townsend Yates[157]

---

[156] Albert and Sarah Anne Webster, Jennie's in-laws.
[157] The next day, Saturday, December 10, 1892, the *Willie Reed* docked at pier forty-six on the East River, New York. Georgia, Jack, and Dorothy Yates had been on board for fifteen and a half months. Eleanor Townsend arrived within a few days to collect her daughter, son-in-law, and granddaughter. The family soon departed for Sycamore and came within sixty miles before Georgia suddenly went into labor. On December 22, 1892, in Chicago's Sherman House hotel, she gave birth to twin girls, Marjorie and Margaret Yates.

Part Three: Life After the Letters

Georgia and Jack in early 1894. (Wendy Jones Smoke Collection)

Sycamore, Illinois

When the *Willie Reed* docked at pier forty-six on New York's East River in December 1892, the press was there, clamoring for information on the mutiny. Charles Schneider, one of the sailors who had been shot and then discharged in Japan, had arrived back in New York two months earlier and had already told his side of the story. The newspaper accounts, which ran nationwide, portrayed Schneider as the hapless victim and Captain Jack Yates as the tyrant. Now, after two months of hearing only a one-sided account, everybody wanted to know Jack's version. In his interview, which appeared the day after his arrival in newspapers from Boston to Nebraska, he described the voyage as "one of the most exciting he ever made," but also "not a pleasant one."[1] His version of events (see pages 89–91) was more detailed and thrilling than Schneider's and reached a much wider audience. Whatever the hearing process may have been—if any hearings were held at all—no charges were ever filed against him. Jack had little time to waste on reporters and investigations, anyway. He had cargo to discharge, a follow-up voyage to arrange, and a wife due to give birth at any moment.

Upon arriving, Georgia sent a telegraph to her mother in Sycamore, who promptly boarded a train and headed east. Within a few days, she collected her daughter, granddaughter, and son-in-law, and they all began the journey back to Sycamore. Georgia's letter of July 24, 1892, indicated that she would not be returning to Sycamore immediately, but would stay in New York or Boston until a few months after giving birth. Perhaps Eleanor convinced her otherwise, or perhaps Georgia decided she'd rather give birth at the family home, as she had done with Dorothy. Interestingly, the Ladies Society of Sycamore's Universalist Church, of which Georgia was a member, hosted a Japanese-themed holiday social a week after Georgia's arrival in New York; perhaps they had planned it in Georgia's honor. Whatever the case, the whole family boarded a train, Sycamore bound. They made it as far as Chicago when Georgia suddenly went into labor.[2]

This was a chaotic time to be in Chicago, giving birth or not. The 1893 Chicago World's Fair – or the Chicago Columbian Exposition, as it was known, because it celebrated the four-hundredth anniversary of Columbus's landing—would be opening in six months. Thousands of contractors packed the city, rushing to finish the nearly two hundred buildings that

---
[1] *New York Press*, Dec. 11, 1892; *Brooklyn Daily Eagle*, Dec. 11, 1892.
[2] The December 29, 1892, issue of the *Sycamore City Weekly* claimed they were trying to make it to Eleanor's in time for Christmas.

covered the sprawling six-hundred-acre fairground. Newspapers claimed it would be the "most significant and greatest exhibition ever seen by man."[3] The restless anticipation brought thousands of visitors to the city each day just to witness the construction of this architectural wonder.

Despite the chaos, Jack and Georgia found lodging at the Sherman House, one of Chicago's finest hotels.[4] On Thursday, December 22, 1892, in room 267, Georgia gave birth to twins, Marjorie and Margaret Yates (known affectionately as Peg and Meg).

The following week, an announcement of the *Willie Reed's* return and the twins' birth ran in the *True Republican*:

> Capt. John E. Yates has been undergoing some thrilling experiences. In company with his courageous wife the Captain went on a cruise to Japan, and during the voyage, which lasted more than a year, they experienced a mutiny with an accompayment [sic] of shooting. The mutiny of the crew was finally quelled, however, and the ship arrived safely in Philadelphia two weeks ago, when Mrs. Yates, accompanied by her mother, Mrs. Townsend, started for their home in Sycamore. Last Friday morning, when at Sherman house, in Chicago, Mrs. Yates gave birth to twin baby girls. Now everybody is feeling well, including the undaunted Captain.[5]

On Monday, December 26, four days after his daughters' birth, Jack returned to New York to attend to business and the unloading of his ship. At this time, Jack sold—or at least arranged to sell—his interest in the *Willie Reed*. By disposing of his share of his most lucrative ship, Jack was also giving up his life at sea. There is no record of him captaining a vessel after this voyage, and a later account of his life mentions a trip to Japan as his last journey on the high seas.[6]

Trying to determine why Jack retired from the sea leads to more speculation. He may have decided before departure that this voyage would be his last, which can be explained by his real estate investments in the West, but it is possible that the events of the voyage pushed him to this decision. First, there was the mutiny, which put his entire crew and the lives of his wife and child in danger. This danger, and its

---

[3] *True Republican*, July 22, 1893.
[4] Opened in 1873, the Sherman House stood at the corner of Clark and Randolph until it was demolished in 1911. It was located on the present-day site of the Thompson Center, 100 West Randolph Street.
[5] *True Republican*, Dec. 28, 1892. The article erroneously reported the ship as arriving in Philadelphia instead of New York. The paper also erroneously reported the twins' day of birth as a Friday. December 22, 1892, the day that appears on the twins' birth records, was a Thursday.
[6] *Idaho Daily Statesman*, Nov. 4, 1900.

Marjorie and Margaret Yates were born in Chicago on December 22, 1892. On Sunday morning, April 22, 1894, they were dedicated at the Sycamore Universalist Church. The local paper noted that "they looked very sweet and attractive as they were presented at the altar by their mother." *True Republican,* May 9, 1894. (Joiner History Room)

accompanying anxiety, would have checked the most seasoned sea captain. Second, the voyage was exceedingly difficult; in addition to the mutiny, it was beset by numerous storms, which kept the captain up long hours, often days at a time. Not only did this prevent him from seeing his family, but it also delayed the trip several months. Third, Jack had his first experience losing a sailor at sea due to disease, which happened slowly and painfully over several weeks. He and Georgia had personally attended the sick sailor and had watched him suffer a terrible death. This would have been a traumatic experience, and Jack certainly worried about the disease spreading to the rest of his crew and, more importantly, to his wife and infant daughter. Finally, Georgia's letters imply that she did not enjoy the trip and most likely would not accompany Jack on another. Her presence on future voyages seemed even less likely with three small children. So if Jack wanted to return to the sea, he knew he would have to do it alone.

While these theories sound promising, we also must consider the state of the shipping industry in the late nineteenth century and how it may have swayed Jack's decision. As early as the 1860s, steam began to encroach into an industry dominated by sail. By the 1890s, the transition from sail to steam – and also from timber to steel – was in full swing. The economic advantages of sailing had passed, and with them went the romanticism, the excitement that was born into Jack's blood. There was still money to be made with old wood sailing vessels, to be sure, but they could never compete with the emerging technologies. With his keen business sense, Jack would have been aware of these shifting tides. Perhaps he had no desire to compete. There is no record of him or his brother investing in a steam ship.

So Jack knew his days at sea were numbered. Just after Dorothy's birth and months before the voyage to Japan, he had traveled west and purchased land in Boise, Idaho, where he and his family would later settle. Disposing of his share of the *Willie Reed* and moving west might have been his plan all along.

On Thursday, January 5, 1893, at a time when "a mantle of snow" covered the land, Georgia and her two-week-old daughters arrived in Sycamore. They retired to Eleanor's "hospitable home," which was "now enlivened by no less than a quartette of wideawake infants in long clothes."[7] Jack finished his business in New York and returned to Sycamore in late February, just in time to attend a large social gathering at the home of George and Sarah Wild, Georgia's aunt and uncle. But he didn't stick around for long. By the middle of March, he was on his way back to Boise, looking for more land to purchase.

At the end of March, Eleanor threw a party at her home. The *True Republican* reported that "the evening was passed in viewing the many interesting relics brought from Mrs. Yates's trip around Cape Horn to Japan."[8] Jack did not return to Sycamore until June. The family then remained with Eleanor for several months. The accommodations were ideal: a large home with modern conveniences and plenty of family and servants to watch over the children.

In late September, Jack's brother and sister-in-law, Oscar and Delia Yates, paid a pleasant visit to Sycamore, where they stayed as Eleanor's guests. Like Jack, Oscar had dedicated his life to the sea; he claimed that this trip to Sycamore was the first time he'd traveled more than fifty miles

---

[7] *True Republican*, Jan. 7, 1893. The other two infants were Georgia's nephews Charles Boynton Townsend, born January 1, 1892, and Edward Denman Claycomb, born March 20, 1892.

[8] *True Republican*, Apr. 1, 1893. Georgia actually sailed around the Cape of Good Hope at the southern tip of Africa. Cape Horn is the southern tip of South America.

from salt water. Oscar might have devoted part of the trip to business and Jack's disposal of his share of the *Willie Reed*, but it is certain that Oscar and Delia also made the trip to visit the Chicago World's Fair, which had opened in May 1893 and was drawing millions of visitors from all over the world.

During the first month, the fair saw over one hundred thousand visitors each day. As it progressed, this number grew to over three hundred thousand. Hundreds of Sycamoreans made the journey each week. In the six months the fair was open (May to October, 1893), it drew over twenty-seven million visitors. Jack, Georgia, and Georgia's family made repeated visits to the fair. On one occasion, Georgia and her cousin Hattie Townsend—daughter of Jack's real estate partner William Townsend—escorted their eighty-four-year-old grandmother, Ann Denman Townsend, to the fair. The *True Republican* noted that Grandmother Townsend "enjoyed the trip and thoroughly appreciated the beauties and wonders of the great Exposition."[9]

In January 1894, Jack hosted a social event at Eleanor's home that was exclusive to his "gentlemen friends."[10] The guest list was a who's-who of Sycamore's business and public elite. The newspaper gave the following account of the "novel" event:

> The party was given by Capt. John E. Yates, and his tact and the solicitude with which he looked after the pleasure of each guest proved that he could guide a craft of this kind just as successfully as he can manage a ship's crew in an Indian Ocean hurricane.

The paper claimed the evening's only interruption was when the twins were brought down "by request."[11]

On February 12, 1894, fourteen months after Jack and Georgia's safe return home, the *Willie Reed* sank in a heavy storm off the coast of France. All twenty-one crew members were plucked from the sea unharmed. After Jack and Georgia's voyage, the ship had departed for Portland, Oregon, under a Captain Forbes. From there it took a shipment of wheat to Ireland and then sailed to Ipswich, England, where it filled its hold for the return to the United States. It departed Ipswich on February 8, sailed through the English Channel, and then foundered off the north coast of France.[12]

---

[9] *True Republican,* Sept. 30, 1893.
[10] *True Republican,* Jan. 20, 1894.
[11] *True Republican,* Jan. 27, 1894.
[12] The *True Republican* reported on the loss on March 17. Multiple American newspapers also reported the wreck of the *Willie Reed*. While many stated that it sank at sea, others said it ran aground. It is possible that it simply foundered at sea and then ran aground later. Either way, it was a total loss.

How Jack and Georgia reacted to this news is unknown, but Jack could take solace that he no longer held a stake in the ship. Oscar had likely also disposed of his shares, because since the day the *Willie Reed* first launched, either Jack or Oscar had stood at the helm. The only time the ship ever sailed under a different captain—Captain Forbes—it sank.

Return to Round Pond

In April 1894, Jack returned to Round Pond to prepare the family home for his wife and three children. Georgia had not been back there since before the voyage. A few weeks later, she and the children, along with Eleanor and Georgia's sister Mary, boarded a train and headed east. Eleanor and Mary stayed for the summer and returned to Sycamore in October. From this point forward, summer excursions to Round Pond became a Townsend family tradition.

Jack and Georgia's fourth child, John Townsend Yates, named for his father, was born in Round Pond on January 1, 1895. From the beginning, he was a sickly child. Eleanor rushed out to assist her daughter in his care, but she did not arrive in time. John Yates died on Friday, March 8. Two days later, he was buried in Chamberlain Cemetery in Round Pond. To compound the tragedy, one day after the funeral, Jack's mother, Sophia Blunt Yates, died suddenly at the age of seventy-five.

Overshadowed by these two misfortunes was Jack's first foray into politics. At the same time as he was burying his son and mother, he was elected as Bristol's town auditor, having run on the Republican ticket. This political venture was another sign that he had put his days at sea behind him.

Over the next few years, Jack and Georgia remained in Round Pond while various family members trickled in for extended summer visits. On February 24, 1896, Georgia gave birth to Oscar Townsend Yates, named after Jack's brother. Eleanor, Georgia's brother, Fred, and his wife, Mary, visited in July of that year. (Fred could stay only a few days because he was now the mayor of Sycamore.) When Eleanor returned to Sycamore in September, Georgia and her four children accompanied her. Georgia spent her time there visiting friends and family before returning to Round Pond in early 1897.

By this point, Jack was planning to move the family to Idaho (in fact, he may have been planning this move since his first trip there in 1891). The death of his mother may have solidified those plans. His brother was also leaving Round Pond. In 1897, Oscar and Delia moved to

Boothbay Harbor, just down the coast. Oscar purchased and operated the Menawarmet, a combination hotel and pool hall. A travel piece in a Maine newspaper described him as "a gentlemanly and obliging landlord...who commands [the Menawarmet] with the same grace and thoroughness as he did his fine ships at sea."[13]

In 1896 and 1897, Jack ran ads in several New England newspapers for his Round Pond home:

> **Summer Home For Sale.**
> Situated on the southern point of Round Pond Harbor, Bristol, Maine. 14 acres land; ten-room house; cemented cellar; furniture, mattresses, carpets, curtains, crockery, cooking utensils; stable 28x30; building at boat landing, 18x20; good bathing beach; land forms a point; has larger water front and numerous cottage sites. Price, $2,200.00.[14]

We don't know for sure why Jack and Georgia decided to move their family to Boise, but the decision may have been influenced by the uncertain political atmosphere that surrounded the American shipping industry. In the mid-1890s, Cuba struggled for independence from Spain, which disrupted shipping to several ports commonly used by the Yates brothers and their partners. This conflict led to American tensions with Spain. Everyone was aware that a war would spread the shipping disruptions worldwide. Contemporary newspapers warned that the Spanish Armada could appear off the New England coast at any moment and bombard America's major port cities. With war seemingly inevitable, the American Navy started paying top dollar to purchase private and commercial ships to outfit for battle.

How this state of affairs affected Captain Yates and his family is un-

This ad for Oscar Yates's hotel ran in the July 8, 1897, issue of the *Brooklyn Daily Eagle*. (author's collection)

---

[13] *Maine Farmer*, Aug. 5, 1897.
[14] *Outlook* 55, no. 17 (Apr. 24, 1897): xi; *Boston Herald*, May 30, 1896. The earliest advertisement mentioned a twenty-one-foot sloop boat with new sails, but later ads didn't include it.

known, but around this time he was liquidating his assets in the East and preparing to remove to the West. The United States declared war on Spain on April 25, 1898. That summer, with tensions running high along the eastern seaboard, Eleanor visited her daughter in Round Pond and helped her prepare for the move. Jack and Georgia and their four children first went to Sycamore, arriving in early September. Jack stayed in town for a couple weeks before heading out to Boise to prepare a home for his family. Georgia and the children followed him in mid-December.

Even though Jack and Georgia had departed Round Pond, Eleanor was not ready to say goodbye to the little seaside community. She purchased a summer cottage and traveled there every year with a rotating mix of her children and grandchildren. One grandson, Amos Claycomb (son of Anna Townsend Claycomb), documented his 1902 trip to the cottage in his

Jack and his four children in Round Pond just prior to the move west. Left to right: the twins, Dorothy, and Oscar. (Joiner History Room)

LIFE AFTER THE LETTERS

This family portrait, depicting four generations, was taken in 1898 while Georgia and her children visited Sycamore before traveling to their new home in Boise, Idaho. (Joiner History Room)

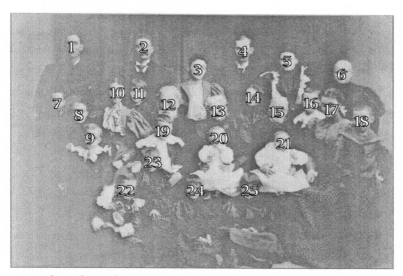

1. Frank E. Claycomb 2. Frederick B. Townsend 3. Mary Boynton Townsend 4. Charles Webster 5. Georgia Townsend Yates 6. Angeline Allen 7. Amos Claycomb 8. Jennie Townsend Webster 9. Pierce Webster 10. Eleanor Claycomb 11. Louise Claycomb 12. Daniel Pierce 13. Eleanor Pierce Townsend 14. Marion Webster 15. Ann Denman Townsend 16. Frederick Webster 17. Mary Townsend 18. George Claycomb 19. Charles Townsend 20. Eleanor Townsend 21. Oscar Yates 22. Edward Claycomb 23. Margaret Yates 24. Marjorie Yates 25. Dorothy Yates

After Georgia departed Round Pond, her mother purchased this summer cottage there and made it an annual destination for the Townsend family. Some family members believe that Eleanor's summer cottage was Jack and Georgia's home, which she purchased when they moved west. This is possible, but has not been confirmed. Fred Townsend took this photograph in September 1903. (Joiner History Room)

journal (see Appendix E). Another grandson, Charles Boynton Townsend (son of Fred Townsend), returned to Round Pond in 1961 to visit the place he remembered so fondly from his youth.

Boise, Idaho

The Boise River cuts a deep channel through the high desert of western Idaho. Its waters nourish a lush green valley that brings life to an otherwise harsh and arid environment. In 1834, the Hudson Bay Company established Fort Boise as a trading post on the river near the present-day Oregon border. The fort dealt mostly in furs and provided a safe haven for frontiersmen and western emigrants. It survived, in one incarnation or another, into the 1850s. The United States established a second Fort Boise in 1863, the site chosen for its strategic location along the Oregon Trail and other popular western routes. Boise City grew around the fort, and within a year the frontier town became the capitol of the Idaho Territory. It remained the capitol when Idaho became the forty-third state in 1890.

In March 1891—around the time Jack first visited Boise—the local newspaper, the *Idaho Daily Statesman,* made the following prediction: "In a few short years these plains now barren, will be rich with the products of the farm, supporting thousands of families, all of whom must look to this city for their necessary supplies and their necessary luxuries."[15] The paper couldn't have been more prescient. Perhaps Jack read that article while in town. Perhaps it inspired him. When he and his family arrived in 1898, Boise was a progressive and fast-growing mining and agricultural center. Its population had doubled since Jack's first visit.

The moment Georgia and the children stepped off the train in downtown Boise, Jack had a new house waiting for them at 1316 N. Eleventh Street. The house had just been finished at a cost of $1,500. It lay north of the city center, in the area now known as the North End neighborhood. Jack held large tracts of land nearby that he would soon use for his many future enterprises.

Following in the footsteps of Georgia's entrepreneurial grandfather, Jack got into the land speculation business. His company, Bogart & Yates, started buying up real estate in and around Boise. Jack's congenial character, coupled with his jovial disposition, made him an intriguing figure around town. The people of Boise found it unique and charming that a globe-trotting sea captain had dropped anchor in their frontier community. He capitalized on his former profession, never missing an opportunity to spin tales of his adventures roaming the high seas or exploring the world's busiest ports. Professionally, he was called Captain Yates, but his friends knew him as Captain Jack. In addition to his real estate ventures, he used his favorable reputation to return to public service. In June 1899, he ran for a seat on the Boise City Council and won easily.

Georgia remained in regular contact with her relatives in Sycamore but did not receive her first visitor, Eleanor, until the summer of 1900. Her mother might have visited sooner, but the previous year she had embarked on a European tour with her daughter Mary and family friend Lucetta (Stark) Boynton, mother-in-law to Fred Townsend. While staying at a hotel in London, all the members of their traveling party fell ill from severe food poisoning, which took the life of their tour guide. The family eventually recovered and returned home. Soon after, Mary married William R. Tapper, whom she had met at Lombard University. In July 1900, the newlyweds traveled to Boise, joining Eleanor there at the Yates's family home.

Jack's ambitions grew alongside his local popularity. By 1902, he was a founding director of both the Boise Bank of Commerce and the Guarantee

[15] *Idaho Daily Statesman,* Mar. 25, 1891.

Title and Trust Company. He founded a fruit-shipping business, Eastman & Yates, which specialized in prunes. He also founded a livestock-trading company, Yates & Corbus, which specialized in sheep.[16] For the former, he joined the executive committee of the South Idaho Fruit Growers Association. For the latter, he joined the executive committee of the Idaho Wool Growers Association. He also became a member of the Idaho Colonization Company, a committee of local businessmen dedicated to bringing new settlers to Boise.

Jack spent a considerable amount of time on horseback, roving the state to further his and Boise's business interests. Due to his wide-ranging travels, the local newspapers often relied on his expertise for reports on crops, livestock, and business in the surrounding region.

While serving on the city council, Jack ran for and lost a spot on the Boise Board of Education. But soon after his defeat he won a seat as a representative in the Idaho State Legislature. During the election, the *Idaho Daily Statesman* threw its full support behind Jack and ran a flattering profile:

> John E. Yates, one of the republican candidates for the legislature, will sail into the house of representatives as easily as, in the old days, when he was master of a ship, he sailed into port. Mr. Yates followed the sea for a quarter of a century, having circumnavigated the globe several times. He took to salt water as naturally as a duck to a pond and he found the transformation to "landlubber" just as easy. Mr. Yates went to sea when a boy of 16. He went up through the various position[s] on ship-board until he became master. The last ship of which he had charge was the famous *William Reed* and his last trip was to Japan. Mr. Yates went to Illinois, where he married and foreswore his love for the briny deep.
>
> Mr. Yates became identified with Idaho for the first time in 1891 and then merely by way of investment. More than satisfied with the investments he had made, he decided to make Idaho his home and move to Boise, where he built himself a beautiful home. Having early identified himself with the progress of the state and having entrenched himself in the confidence and esteem of his fellow-men, he was soon looked upon as one of the most substantial citizens of the county. His popularity was attested and the confidence reposed in him demonstrated by his election to the city council, in which capacity he has served the city well, being ever in the forefront in the prosecution of any move looking to the advancement of the municipality's best interests.[17]

---

[16] Yates & Corbus was a partnership between Jack and twenty-six-year-old Fred Corbus, a native of Boise. The partnership was short-lived; Fred Corbus was killed in a horse-riding accident within a year of the company's incorporation.

[17] *Idaho Daily Statesman*, Nov. 4, 1900.

Not one to stand in her husband's shadow, Georgia threw herself into Boise's busy social scene. She became a fixture at various parties and charity events. She served on the local library committee and became a member of the local branch of the Columbian Club, a national women's organization that promoted education. Georgia once gave a talk to the club on the Consumer's League, a new group dedicated to improving the working conditions of women. At some point, Georgia left the Universalist denomination and moved to its close cousin, Unitarianism. (The two denominations later combined in 1961). In 1901, she co-founded a Unitarian church, the Unity Church of Boise. Thenceforward, she devoted much of her free time to church initiatives. Unitarian church records list her as the church's "chairman or clerk" in 1901 and 1910. From 1901 to 1903, she served as the vice president of the Boise branch of the National Alliance of Unitarian and Other Liberal Christian Women.

Amidst the Yates family's flurry of activity, Georgia gave birth to a sixth child, Frederick Townsend Yates, on May 13, 1901. He was named after Georgia's brother.

The Yates family rode high on a tide of prosperity, but over the next few years, multiple tragedies struck. Georgia's grandfather, Daniel Pierce, who had migrated from New York to Sycamore to secure the family's fortune, passed away on April 27, 1902, at the age of eighty-seven. Two weeks before his death, Georgia's mother and sister Jennie had traveled to visit Georgia in Boise. They no doubt wanted to see little Fred, the newest addition to the family. While there, Eleanor received word that her father was sick. She immediately left for home, but he passed away before her return. At the time of his death, Daniel Pierce had been mostly retired and in poor health for several years. He left an estate valued at $1.5 million.

That summer, Eleanor returned to her seaside cottage in Round Pond and took along several of her grandchildren.[18] During their stay, they received the news that Angeline Allen, who had worked in the Townsend home for over forty years and helped raise Georgia and all her siblings, had passed away on July 15. On July 30, they received a telegram that Mary, the youngest Townsend sibling, was in poor health. She had given birth recently and never quite recovered. The family hastily departed Round Pond and reached Chicago on August 1 to be at Mary's side. She passed away on August 7, just a few weeks shy of her twenty-seventh birthday. Six

---

[18] This trip to Round Pond was recorded in the diary of Georgia's nephew Amos Claycomb and is included in Appendix D.

Georgia's younger sister, Mary Townsend Tapper, passed away on August 7, 1902, just shy of her twenty-seventh birthday. (Joiner History Room)

weeks later, on September 22, Georgia's ninety-three-year-old grandmother, Ann Denman Townsend, also passed.

For the first time in her life, Eleanor Townsend occupied an empty house. To alleviate this condition, she spent several weeks with her daughter Jennie in Galesburg in May 1903. Georgia traveled from Boise with all five of her children to meet them there. Georgia then accompanied her mother back to Sycamore and stayed with her for several weeks. In that time, the Townsends enjoyed a large and festive family gathering to celebrate the Fourth of July. In late July, Eleanor, Georgia, and the five children returned to the seaside cottage at Round Pond. It was the first time Georgia had visited since departing five years earlier, and it must have been an

Four of the five Yates children who traveled back to Sycamore and Round Pond in the summer of 1903. Left to right: Dorothy, Margaret, Marjorie, and Oscar. Not pictured: two-year-old Fred. (Wendy Jones Smoke Collection)

exciting homecoming for twelve-year-old Dorothy and the ten-year-old twins. Georgia and her children stayed for several weeks before returning to Boise.

Upon their return, Jack purchased a thirty-two-acre fruit ranch about two miles west of Boise, near his flourishing orchard operations. For some time, Jack had been growing, drying, and packing prunes on land just west of the Idaho Soldier's Home (present-day Veterans Memorial Park). His new purchase stood on the east side of the Soldier's Home, just off Valley Road (present-day West State Street). The spring-fed property included a large and elegant brick home, one of Boise's oldest. Rancher George Davis, one of Boise's earliest settlers, had built the house thirty years before. Jack and Georgia relocated to the new home, which would become the site of several premier Boise social events. They named the ranch "The Anchorage," a fitting title for the dwelling of an old sea captain. Jack and Georgia Yates were anchoring their lives to this ranch in Boise, Idaho, because it was secure, a source of

This image of Jack (arrow is pointing to his foot) and unknown members of a party was included in the scrapbook of Georgia's brother, Fred Townsend. Surrounding it were pictures of Round Pond taken in the summer of 1903. So we can assume that Jack traveled with Georgia and the children that summer and this picture was taken there, possibly with old friends or family. (Joiner History Room)

The Idaho Soldier's Home opened in 1895 as a place to care for aging, poor, or disabled veterans. It occupied forty acres that rancher George Davis sold to the state for $5,000. The Yates's new property, also bought from George Davis, was located to the east of this building. The onion-shaped dome was added after a 1901 fire and would have been highly visible from the Yates's home. (Idaho State Historical Society 76-114-8c)

Jack and Georgia Yates standing outside their ranch home, The Anchorage. (Wendy Jones Smoke Collection)

reassurance, the place they planned to raise their children and spend the rest of their lives.

On October 1, 1904, The Anchorage welcomed the birth of Jack and Georgia's seventh child, William Townsend Yates. While there were several Williams in Georgia's family, he was most likely named for her uncle William H. Townsend, who had joined Jack on his first land speculation trip to the West.

Jack and Georgia's joy, however, was short lived. Georgia's mother, Eleanor Pierce Townsend, passed away on December 20, 1904, at sixty-five years old. She had been sick for several months, but Georgia had not been able to visit due to her pregnancy. Many family members believed that Eleanor had never fully recovered from the food poisoning she had suffered in England five years earlier.

Her obituary in Sycamore's *True Republican* noted that "seldom has an event caused more general sorrow in this community than the death of Mrs. Eleanor P. Townsend." It described her as "a lady of much energy, interested and active not only in the welfare of all of her many relatives, …[but] one of the most generous contributors of her means and her sympathies to every public benefaction and every cause for the benefit of the people of whatever degree." Several friends and associates expressed their condolences in the local paper, with one writing: "To the friends who have known her intimately no sadder words can ever be uttered than those heard on our streets last Wednesday morning—'Mrs. Townsend has gone.'"[19]

At the time of her death, she had eighteen grandchildren and adored them all, which she proved with her constant travels to visit them or by enlisting them as her travel companions. In her will, she left substantial sums to Lombard University and Sycamore's Universalist Church.

Without any personal writings from Georgia during this time, we have no evidence of how the loss of her mother affected her. Over the next few years, the heirs to the Pierce estate – Fred, Jennie, Georgia, and Anna and Mary's families – divided and sold most of their property in and around DeKalb County. Jack and Georgia sold most of their inheritance. Jack made several trips to Sycamore to handle the transactions personally. Frederick Townsend, the only Townsend sibling still residing in Sycamore, purchased and consolidated most of the original estate.

In 1909, Eleanor Townsend's surviving children—Fred, Jennie, and Georgia—donated the Townsend Mansion on Somonauk Street for use as a hospital and old people's home. The *True Republican* noted:

> It is eminently fitting that the home of one of Sycamore's noblest women should be devoted to the cause she loved and supported while she dwelt among us. Mrs. Townsend was beloved by all who knew her and her wealth was always at the disposal of the poor and needy. Her charity was boundless, and her love for suffering humanity led her to consecrate a large portion of her means towards relieving their wants.[20]

---

[19] *True Republican*, Dec. 24, 1904.
[20] *True Republican*, Jan. 29, 1910.

The Sycamore Memorial Hospital opened in 1912. A new wing was added to the mansion in 1917. Major additions followed in 1949 and 1958, but the hospital demolished the original mansion section in March 1962 to make way for a modern facility. The hospital still operates today as Kindred Hospital Sycamore.

In 1906, Jack was busier than ever. In January, he helped create the Boise City Canal Company, which constructed bridges within the city. In March, he bought a controlling share of the Bank of Commerce (which he had cofounded) and became its president. He continued buying and selling real estate and in May began planning a new downtown office building. The one-story, $4,500 structure would join several other buildings Jack already owned near Boise's railroad station. Jack also served as the chair of the local Republican committee and formed a commercial club dedicated to bringing better roads to Boise (modeled after, so he claimed, the fine paved roads he found in Sycamore). In November, he co-founded the Farmers' Bank of Star in the town of Star, fifteen miles northwest of Boise, and in December, he cofounded the Idaho Title & Trust Company.

Georgia remained busy with her church and social activities. Her sister Jennie visited so often that she also became a social staple in Boise.

In 1909, the Townsend family donated Eleanor's home on Somonauk Street for use as the Sycamore Memorial Hospital. This postcard image shows a hospital wing extending off the original structure. The hospital removed the mansion section in 1962 to make room for a modern facility. The hospital still operates today as Kindred Hospital Sycamore. (Joiner History Room)

LIFE AFTER THE LETTERS

The Anchorage became the site for several meetings and events. Jack held political gatherings there, and Georgia hosted church and civic organizations. Because the ranch was up the valley outside of Boise, large carry-all wagons shuttled the guests from downtown. In one event of note, members of one of Georgia's social clubs gathered at The Anchorage in June 1906 for an evening of several unspecified games. As prizes, she offered souvenirs from her voyage to Japan. The society page of the June 24 issue of the *Idaho Daily Statesman* noted that Georgia gave away her Chinese and Japanese teapots as well as a set of Chinese tea cups.

At the close of 1906, Jack and Georgia sold thirty-two acres of land between their home and the Soldier's Home to be used as a public park. The Boise Traction Company purchased the land, extending its streetcar line to the park so guests could enjoy picnics and other public events along the Boise River. The company called it Yates Park, and it operated until 1917. It is now part of Veterans Memorial Park.

Jack and Georgia's youngest child, Stephen Townsend Yates, was born November 26, 1907. Jack was sixty-two years old and Georgia was forty-one. Stephen was named for Georgia's paternal grandfather.

Soon after Stephen's birth, Jack completed construction on his downtown building, which rented office space to various Boise ventures. In January 1909, the Idaho Trust and Savings Bank bought out and absorbed Jack's Bank of Commerce and took immediate control of

All four Yates boys on the Boise ranch. Left to right: Oscar, Fred, Bill, and Stephen. (Wendy Jones Smoke Collection)

Built in 1905, the Yates Building took its name after Jack assumed ownership in a 1909 bank merger. This prominent downtown building was located on the southwest corner of Ninth and Main and offered prime office and retail space. It was home to Alexander's, a popular Boise clothing store. (Idaho State Historical Society 74-105-24)

operations. It retained Jack for its board of directors. It also moved its main operations into the Bank of Commerce building, which had been fully renovated three years before. As part of the buyout deal, Jack and his business associates assumed ownership of the large office building at Ninth and Main that the Idaho Trust and Savings had vacated. Noted Boise architect J. E. Tourtellotte had constructed the building in 1905. Most of its offices already had tenants, including Alexander's, a popular clothing store that had been a fixture in Boise for over twenty years. Within a week of acquiring the building, Jack leased the vacated bank space to a rival, the Boise State Bank. Soon after, the building became known as the Yates Building. It was Jack's most valuable piece of real estate.

This success was overshadowed when a fire destroyed or damaged several properties that Jack owned. Always one to seek new opportunities no matter how they presented themselves, Jack vowed to rebuild bigger and better than before. In 1910, he broke ground on his most ambitious project yet, the Bristol Hotel, named in honor of his birthplace. Designed by architects Nisbet & Paradice, the $50,000 four-story building stood only a block from Boise's railroad depot and incorporated the most

LIFE AFTER THE LETTERS

modern amenities. The *Idaho Daily Statesman* provided an overview:

> The entire building will be heated and lighted in the latest manner, with ventilation. Every room will be an outside one and one-half the rooms will be equipped with private baths, aside from the general baths. An automatic self-elevator will be installed, also telephones, built-in wardrobes, hot and cold water in each room. In fact, all of the latest hotel equipment.[21]

The hotel opened in mid-February 1911. While the building was under construction, however, Jack had decided to take another run at public office. He hadn't held a public position since his 1900 to 1902 term as a state representative, but he had always remained politically active. In November 1910, he ran as the Republican candidate for state senate and won. One of his first duties was to head the Committee on Banks and Banking. The following year he helped plan President Taft's campaign visit to Boise.

In June 1912, Jack traveled to Chicago to attend the Republican National Convention. While there, a little nostalgia overtook him. He visited the Sherman House and requested to stay in room 267. When told the room was occupied, Jack persisted, explaining that it was the room

This postcard image shows Jack's biggest business achievement, the Bristol Hotel at the corner of Tenth and Grove. Jack named it in honor of his hometown in Maine. When it opened in 1911, it was Boise's largest hotel and benefited from its close proximity to the railroad depot. (author's collection)

---

[21] *Idaho Daily Statesman,* Mar. 17, 1910.

his twin girls had been born in almost twenty years earlier. The hotel staff made arrangements with the room's occupant and honored Jack's wish.[22]

The 1912 Republican National Convention became infamous for its bitter division of the Republican Party. Republicans had to choose between nominating the incumbent president, William Howard Taft, or making the unprecedented move of ditching the incumbent and nominating former president Theodore Roosevelt, who sought a third term. Jack threw his support behind Roosevelt. At the tense and heated convention, Taft secured the nomination. However, bitter lines had been drawn and irreparable damage done. Roosevelt's supporters cried foul and claimed that the nomination was stolen. Jack told a paper in Boise that he "knows and believes full well that Taft was not the honest choice of that convention and that he was not nominated and will so declare."[23] Roosevelt and his supporters – Jack included – stormed out of the convention, organized a new convention of their own, and created a new political party: the Progressive Party (popularly known as the Bull Moose Party).

When Jack returned to Idaho, the first-term Republican state senator declared his allegiance to the Progressive Party and announced that he would run for state treasurer under the new party's banner. He immediately organized a Progressive movement in Boise by calling a meeting of prominent citizens where "plans were laid for an active and aggressive campaign of education along Progressive lines in Idaho from now on."[24] He chaired a committee that brought Roosevelt to town in September. Thereafter, he canvassed the state, stumping for Roosevelt, himself, and the new party. He spoke before crowds of hundreds, with Boise's *Evening Capital News* noting after one speech that Jack had "sprung a great surprise on his friends by showing himself a campaign speaker of considerable ability" who "spoke earnestly and effectively and was frequently interrupted with applause."[25]

But the Progressives never gained much of a foothold, nationally or in Idaho. Roosevelt lost his presidential bid just as Jack lost his bid for state treasurer. The Bull Moose Party did manage to split the Republican vote, however, ensuring that Democrat Woodrow Wilson secured the presidency.

---

[22] The Sherman House in which Jack's daughters were born was razed in 1911 to make way for a more modern building, also called the Sherman House. So even though Jack got to stay in a room with the same number as the one his daughters were born in, it was not the same room (or even the same building).
[23] *Evening Capitol News* (Boise, ID), Sept. 16, 1912.
[24] *Salt Lake Tribune*, July 16, 1913.
[25] *Evening Capital News*, Oct. 3, 1912. For a full transcript of a speech Jack gave during this campaign, see Appendix D.

Jack stayed focused on his business and political endeavors while Georgia remained active in her social organizations and took a stronger role in her church. She served as a Unity Church trustee and later became president of the board of directors. She also ran the church's Woman's Alliance Committee and served as president of the Boise Unitarian Society (and toastmaster of its social gatherings).

Around 1909, Georgia became involved with local theater. She started as a patron and then began taking small parts in plays. In a production of *Babes in the Woods*, she played the role of "villain uncle." The local paper gave the following critique: "If Captain Yates were present and was told the strident, blustering uncle on the stage was the wife of his bosom, he certainly would have marveled."[26]

Georgia continued to host annual visits from her sister. She also received regular visits from Sycamore relatives who had moved to Idaho's capital. Two of her cousins (William Townsend's daughters) ended up living in Boise. Georgia's nephew, George Claycomb, moved to Boise in 1910. He stayed with Jack and Georgia while he worked to secure a ranch of his own. George eventually expanded his property to several hundred acres and specialized in raising cattle. A niece, Alta Louise Claycomb, also moved to Boise. In 1911, she married Lewis Ray Love in a ceremony at Jack and Georgia's ranch.

On July 20, 1913, Jack and Georgia celebrated the marriage of their oldest daughter, Dorothy, to Lee Clinton Morehead, of Sutter, California. Dorothy and Lee had met while attending the University of California at Berkeley. They married in Boise and settled in Sutter. Within a few months, Dorothy was pregnant with their first child, also the first grandchild for Jack and Georgia.

Soon after the wedding, Jack and Georgia sent their twin daughters across the country to attend Simmons College, a private women's college in Boston, a city that Jack had visited often in his sea captain days. Along the way, Marjorie and Margaret stopped in Sycamore to visit their aunts and uncles and many cousins.

The family's good fortunes, however, were again marred by tragedy.

On Sunday, March 1, 1914, Jack suffered a mild heart attack while cranking his automobile. For the next two days, his condition steadily worsened and he remained bedridden. On Tuesday, doctors seemed to think he was on the mend. But on Wednesday, March 4, he suffered

[26] *Idaho Daily Statesman*, Apr. 19, 1909.

another heart attack around 6:30 a.m. Georgia left his side to telephone the doctor, and when she returned, he had passed away. He was sixty-nine years old.

The loss of Captain Yates devastated both his family and his community. News of his death spread far and wide, and friends and relatives from all over the country soon arrived at The Anchorage. The *Idaho Daily Statesman* noted that Boise had lost one of its leading citizens of the last twenty years, calling Jack a "pioneer of Idaho" and "one of the builders of Boise."[27] The *Evening Capital News* of Boise wrote, "He was regarded as an able and conscientious officer, who, though he stood alone, would fight to the end on a point which he believed was right." It also noted that had he lived, he would have again run for office under the new Progressive Party.[28] Georgia's hometown newspaper, the *True Republican*, also paid tribute, describing Jack as "genial, friendly and entertaining," someone who "made many friends during the several short periods when he made residence in Sycamore."[29]

A memorial service was held at The Anchorage on Monday, March 9. Captain John E. Yates's casket was then escorted to the train station by members of the Masonic and Elks lodges. It traveled to Portland, Oregon, where Jack's remains were cremated. The final resting place of his ashes remains unknown.

Captain Jack Yates was a restless soul, always changing professions, always evolving. He sought the sea, and then left the sea. He bought ships, and then sold those ships. He bought land, and then sold that, too. He started business after business, built each one up, sold it, and moved on to the next. He pursued public office, sometimes won, but rarely ran for re-election to the same position. Jack was a force of character, beloved as a self-collected man who led by example, inspiring others with his success. Whether it was in business, in public service, or at sea, he remained a wanderer. In the words of Ginny Snyder, his great-granddaughter, "He was always a sailor regardless of whether he was at sea or not. There was always that next frontier."[30]

At forty-seven, Georgia Townsend Yates found herself a widow with four of her seven children still living at home. The youngest, Stephen, was only six. Her first grandchild was also on the way. Like her mother before

---

[27] *Idaho Daily Statesman*, Mar. 5, 1914; Mar. 15, 1914.
[28] *Evening Capital News*, Mar. 4, 1914.
[29] *True Republican*, Mar. 25, 1914.
[30] Ginny Snyder, interview by author. Digital recording. Wenatchee, WA, Sept. 13, 2014.

LIFE AFTER THE LETTERS

John Elvin Yates passed away on March 4, 1914, at sixty-nine years old. Georgia found herself a widow at forty-seven with four of her seven children still living at home. (Wendy Jones Smoke Collection)

her, Georgia now had to stand tall in the face of tragedy and become the central figure in the family. Financially, Jack had left her in comfortable circumstances. His estate was estimated at a quarter-million dollars and included valuable real estate all over Boise. But the loss of her husband was only the first of a series of misfortunes that befell Georgia and her family.

At first, life moved on. On June 3, 1914, Dorothy gave birth to a daughter, Penelope Morehead, at The Anchorage. That same month, Georgia oversaw major renovations to the Yates Building for a new tenant. In October, Georgia's son Oscar was accepted into West Point Military Academy. In November, Georgia visited family in Sycamore before traveling to see her daughter Marjorie in Boston. (Margaret did not return to school after coming home for her father's funeral.) Over the next year, she received numerous visitors at her home, both friends and family. On August 25, 1915, Margaret married Frederick Breitinger of Philadelphia. The small but intimate ceremony took place at The Anchorage.

Then, two weeks later, Dorothy's husband died in a threshing accident. Lee Clinton Morehead was twenty-five years old. At the time, Dorothy was pregnant with their second child. She immediately moved in with her mother at The Anchorage. Her second child, a daughter, was born six weeks later. Dorothy named her Lee Clinton Morehead in honor of the father she would never meet.

But again life moved on. Oscar graduated, enlisted in the military, married the socialite daughter of a prominent Boise family, and shipped off to Europe to serve during the First World War.[31] Marjorie graduated from college and had a rushed wedding when she found out her fiancé, Isaac Clyde Cornog, also was due to deploy. Georgia escorted her daughter from Boise to Virginia so the couple could be married at Fortress Monroe.

Georgia continued working with her social and civic clubs and hosting several events at her home. Like her mother had done after losing her spouse, Georgia spent considerable time traveling the country to visit far-flung family and friends. Those same family and friends also made regular visits to The Anchorage.

Georgia oversaw Jack's business interests and continued buying and selling Boise real estate. But by 1920, she was selling more than buying. She soon put her two largest assets, the Yates Building and the Bristol Hotel, on the market. In July 1921, she received offers on both. Interested buyers wanted to purchase her share of the Yates Building for $80,000 and her share of the Bristol Hotel for $100,000. The sale of the Yates Building eventually fell apart and Georgia retained ownership, but the sale of the Bristol went through for the agreed-upon sum. At the time, the decade-old business was still the newest hotel in downtown Boise.

Georgia with her grandchildren at The Anchorage around 1917. The oldest girl is Dorothy's daughter Penelope Morehead. The other two children are most likely Dorothy's daughter Lee Clinton Morehead and Margaret's son Frederick Breitinger. (Joiner History Room)

---

[31] While serving in France, Oscar was promoted to captain, making him Boise's new Captain Yates.

LIFE AFTER THE LETTERS

Dorothy remarried a man named Alfred R. Halverson in a quiet ceremony at The Anchorage. As with most of the weddings that took place there, it was a small affair limited to family and a few friends. The couple moved into a cottage home Georgia had built for them on the ranch property.

Five months later, on March 8, 1922, The Anchorage burned to the ground.

> ### Boise Landmark Razed by Flames
> Fire of unknown origin Wednesday afternoon completely destroyed the half-century-old "Captain Yates house," a Boise residential landmark of two stories, and brick construction, lying just east of the Soldiers' home. The house, owned and occupied by Mrs. Georgia Yates and her sons, Fred and William, was entirely gutted. . . .
>
> A defective flue is thought to have been the cause of the fire, which gained great headway before its discovery by Fred Yates, a son, who saw flames leaping from the blazing roof on his return home about 3:20 o'clock. The flames continued to spread rapidly over the house, while volunteers from a crowd of approximately 200 persons gathered at the fire were instrumental in saving the household property, family heirlooms and valued and costly mementoes of Mrs. Yates' foreign journeys.[32]

Many of the two hundred volunteers who helped save the household property were employees and tenants of the Soldier's Home next door. They raced over as soon as they saw the smoke. According to a family story, neighbors saved most of the items, including several trunks, by hurriedly tossing them out of doors and windows. So while much of the property was saved from the fire, many valuables were broken in the attempt to save them.[33]

After the fire, Georgia, Fred, and William moved in with Dorothy's family. At the time, the youngest son, fourteen-year-old Stephen, was away attending Swarthmore Preparatory School outside Philadelphia, where both Marjorie and Margaret lived with their husbands and children.

A greater tragedy struck in the summer of 1924. Nineteen-year-old William Yates was on summer break from Lombard College in Galesburg, Illinois. On July 27, while swinging from a trapeze that hung over the pool at the Boise Natatorium, he attempted a complicated dive into the

---

[32] *Idaho Daily Statesman*, Mar. 9, 1922.
[33] Ginny Snyder noted that it was possible that Georgia's letters were in one of these trunks, which was how they survived to be transcribed and passed down to us today.

water. Something went wrong, he dropped too close to the side, and his head struck a pipe that ran around the pool's edge. His mother and two siblings, Dorothy and Stephen, were watching. Dr. Fred Pittinger, a noted Boise physician (and also Oscar's father-in-law) was already at the natatorium and rushed to the young man's aid, only to find that William had died upon impact.

This latest heartbreak was too much for Georgia. Soon after William's death, she deeded the ranch to Dorothy and departed Boise for good. She took Stephen, her youngest son, and relocated to Seattle, Washington.

Seattle, Washington

Georgia's first cousin, Catherine Ann Brundage, had moved to Seattle around 1910. Georgia had made regular visits to see her, so she would have been familiar with the area. Perhaps she had been planning to move there for some time, but it is likely that The Anchorage fire and death of her son precipitated the move. Boise held too many bad memories.

After Georgia departed Boise, she disappeared from much of the historical record. Her life in Sycamore had been well documented because of her family's prominence in that city. She received similar coverage in Boise due to her husband's role in local business and politics. After his death, she controlled his business interests, so her dealings in that regard continued to make the local papers. When Georgia moved to Seattle, however, little record was made of her life or daily activities. Georgia was nearly sixty years old, living in a much larger city, and without the family connections that put one in the daily society pages.

Much of what we know about Georgia from this time period comes via two of her great-granddaughters, Virginia "Ginny" Snyder and Wendy Smoke, the granddaughters of Georgia's son Fred. Ginny, the oldest of Fred's granddaughters, remembers visiting Georgia when she lived in a small cottage behind Fred's Seattle home.

Fred was always an integral part of Georgia's life, but when he was younger, the "wild child" of the Yates siblings often butted heads with his mother. He had inherited both his father's wanderlust and his love of the sea. While his more dutiful older brother, Oscar, served overseas, Fred dropped out of Boise High School, ran off to the Pacific Coast, and joined the crew of the merchant vessel *Phyllis Comyn*. The five-masted barkentine was one of the last large sailboats operating in the age of steam. Perhaps Fred, who grew up hearing stories of his father's ocean adventures, longed to walk in the old man's footsteps. The ship was owned by Comyn, Mackall & Company, a large lumber exporting firm based in San Francisco. The *Phyllis Comyn* sailed up and down the Pacific Coast, loading lumber destined for prominent ports throughout the Pacific, including Australia and South America.

Eventually, Fred returned to Boise, where one of his old teachers encouraged him to finish school. He did so, even though he was older than the other students. During this time, Georgia took it upon herself to find Fred a wife. She picked the socialite daughter of another prominent Boise family.[34] Georgia went so far as to send the girl to a school in Philadelphia so she would receive an "appropriate" education. Fred, however, had no interest in this girl. He had already met someone else.

Virginia Carolyn Stewart was from a working-class family. Not only was she from a different part of town, her parents were divorced. Her mother had fled Iowa with Virginia and her siblings to escape a brutal husband. They had started over in Boise, where Virginia's mother found work as a seamstress. Virginia often assisted her, sacrificing her childhood to help her mother bring home a few extra dollars each week.

While Georgia had come from the upper crust, a stiff-necked world where appearances meant everything and ladies were expected to act a certain way, Virginia's family had struggled to put food on the table. Virginia grew up a tomboy, flouting social conventions, especially those that dictated how women were supposed to behave. Despite her family's disadvantages, she loved reading and learning. When she and Fred met, she was highly educated, extroverted, strongly independent, fiercely loyal to her family, and extremely witty.

Virginia wanted to prove that she wasn't some tramp trying to trap Georgia's son—as her granddaughter Ginny phrased it—but she also would not stand for pretense or fuss. She was unconventional, and Georgia did not approve of anything outside of convention. As a result, these two independent, headstrong women clashed constantly. Several of their descendants have a theory as to why, and it has nothing to do with class or social status; according to those who knew both women, they could never get along because they were so much alike (which could be

---

[34] Oscar Yates had married the socialite daughter of the well-known Pittinger family. Georgia and the bride's mother, Alice Butterworth Pittinger, had been close friends, but it is unknown whether the marriage was arranged. Whatever the story, their marriage eventually ended in a bitter public divorce.

why Fred was attracted to Virginia in the first place).[35]

Even though Georgia made no attempts to hide her disapproval of the relationship, she eventually came to accept it. On February 15, 1923, Fred and Virginia married at Dorothy's home on the Yates ranch. As a gift to her wayward son, Georgia transferred ownership of a one-hundred-acre ranch north of Boise. On the surface, she seemed to be giving her son a fresh start, but she may have had an ulterior motive for this gift: trying to ground her roving, happy-go-lucky son so she could keep tabs on him. Whatever the motive, Fred tried to make a go of it, but was never successful. The farm was a hardscrabble place, not a real gift or a fresh start. In the words of his granddaughter, "He had saltwater in his veins, and he couldn't farm."[36] Virginia made more money selling eggs and homemade butter than Fred did doing anything else on the farm. They knew it couldn't last, and not just because of the money. Fred had already tasted the sea life, and he couldn't wait to get back.

Not long after Georgia left for Seattle, Fred put his ranch up for auction, including all the livestock, machinery, and equipment. He packed up his wife and newborn daughter, whom he'd named Georgia Townsend Yates in his mother's honor, and followed her to Seattle.

Frederick Yates on his ranch north of Boise. (Wendy Jones Smoke Collection)

Tensions between Georgia and Virginia began immediately. Georgia had already picked out a house for Fred within walking distance of her own. Virginia didn't like it and didn't want to live there, but Fred gave in to his mother's wishes and bought the house anyway. It stood at 6252 Twenty-Seventh Avenue NE, and Georgia became a fixture at

---

[35] Despite their differences, it was Virginia's idea to transcribe and preserve Georgia's sea letters for her children and for future generations.
[36] Snyder, interview.

192

LIFE AFTER THE LETTERS

Frederick Yates's house at 6252 Twenty-Seventh Avenue NE, Seattle. Georgia eventually lived in a small cottage that Fred built for her behind the house. (Wendy Jones Smoke Collection)

the dinner table.

Fred found work as an auto mechanic and then later for a gas company. He eventually took a job at a marina, where he repaired and refueled boats. According to Ginny, "There wasn't anything Grandpa didn't know about boats. It didn't matter if it was a rowboat, right on up to managing the electrical systems of a freighter… He must have got a lot of the early stuff from his father."

In his house, Fred kept a room full of sailing memorabilia. On one wall hung a painting of the *Willie Reed*. On a shelf stood a box that displayed his father's compass, which had joined Jack on his travels all over the world.

The youngest Yates sibling, Stephen, finished high school in Seattle. He eventually cofounded a boat dealership and marina in Portage Bay, a spot on the Montlake cut between Lake Washington and Lake Union. The Yates brothers spent their free time on their own boats and became loyal members of various area yacht clubs, which Ginny described as "social extensions of the families."

Fred and Stephen always remained close, as did their children. The brothers eventually started their own marina, Yates Marine Service. Like their father, they dedicated their lives to the sea.

Georgia, however, never took an interest in her sons' maritime

Frederick Yates shared his father's love of the sea. (Wendy Jones Smoke Collection)

Here is Fred pictured with his daughter, Georgia Townsend Yates. Georgia went by "Bridget" to avoid confusion with her grandmother. (Wendy Jones Smoke Collection)

LIFE AFTER THE LETTERS

Yates Marine Service in Seattle, owned and operated by Georgia's sons Fred and Stephen. (Wendy Jones Smoke Collection)

activities. As far as anyone in the family could remember, she never stepped foot on a boat after moving to Seattle.[37]

Georgia had relocated to Seattle with a large fortune, which she had sustained by disposing of the rest of her Boise property piecemeal. In 1929, she sold her most valuable possession, the Yates Building.[38] However, according to family stories, Georgia was never good at managing her money. She put it in the control of lawyers who were not looking out for her best interests. Ginny remembered her grandmother once saying that "all the lawyers fleeced [Georgia] completely." Some family members believed Georgia might have been pushed toward several bad investments. Whether she received bad advice or not, she was clearly investing heavily in the late 1920s, just before the Great Depression. So she may have been swindled out of her fortune, or she may have lost it in the stock market crash that decimated most of the country's investments.

Whatever the story, in the 1930s Georgia lost her money and eventually her home. She moved in with her daughter Dorothy, who had

---

[37] Some family members have questioned whether Georgia ever stepped foot on a boat after the voyage documented in her letters. But there is evidence that she took small trips. According to the July 23, 1909, issue of the *San Juan Islander*, she took a vacation to the island with two of her cousins. But there is no evidence of her taking any other extended voyages.

[38] It is unknown how much Georgia received for the Yates Building. She held the larger of two interests in the property. At the time of the sale, it was valued at $100,000, but that may not have been the final sale price.

Georgia with her grandchildren Georgia and Fred Jr., ca. 1935. (Wendy Jones Smoke Collection)

recently moved to Seattle under her own financial difficulties; Dorothy had lost her home and property in Boise to foreclosure.

Around 1940, when construction materials were being diverted to the war effort, Fred managed to stockpile some scrap lumber behind his house. With the help of a couple friends, he built a small cottage for his mother. Every day after work, before he went inside to see his own family, Fred would walk to the back of the house, sit with his mother, and chat with her about the day.

Ginny Snyder remembers the cottage and her great-grandmother. She recalls that Georgia kept several cats, that she had a beautiful cut-glass candy jar that the children weren't allowed to touch, that she liked to listen to baseball on the radio, and that she would sit outside her cottage in a rocking chair and read or knit:

> She knit on very fine needles with very fine yarn. I mean, it was tiny. And she made baby booties. And more baby booties. And more baby booties. Apparently she just kept making baby booties. I don't know if she was losing it or if this was something to do, I don't know. But they were wool, they were beautifully knit… And those pairs were put in quart jars. She saved them in quart jars because no moths could get in. And she saved them on and on and on. People used to make jokes about it because *if you got a pair of Georgia's booties,* apparently you were very high on the list. And it was almost as if she… almost as if that was some sort of leverage with her. You know, she, by then, was no longer the domineering figure that she had been. And certain things had been a great comedown, and I think a lot of it had to do with money. You know, she ran out, and that was that.

Looking back on her great-grandparents' lives, pondering all that they accomplished during their travels overseas and then after resettling in the West, Ginny noted, "They were pioneers." Reflecting on Georgia, the heartbreak and multiple tragedies she endured after the loss of her husband, the fortitude it took to move forward and hold her family together, she remarked, "She had rebar for a spine, I'm sure."

In a 1951 issue of the *True Republican,* eighty-four-year-old Sycamore historian O. T. Willard mentioned that he and an old classmate from Sycamore High School, Georgia Townsend Yates, now living in Seattle, had exchanged several letters. He also gave her address and mentioned that "she would be happy to receive letters from her friends of long ago." Willard died the following year. His correspondence has not survived. But his one brief mention of Georgia shows that for the rest of her life, she never forgot her roots and remained in affectionate contact with the Sycamore of her youth.

Georgia Townsend Yates passed away on November 17, 1955, in Seattle, Washington. She was eighty-nine years old.

Georgia knitting outside her cottage in Seattle, ca. 1950. (Wendy Jones Smoke Collection)

Appendices

The women in my family are maverick wanderers; as you can see we came by it legitimately.
—Wendy Smoke (Georgia's great-granddaughter)

Appendix A
Family Trees

Georgia's Family

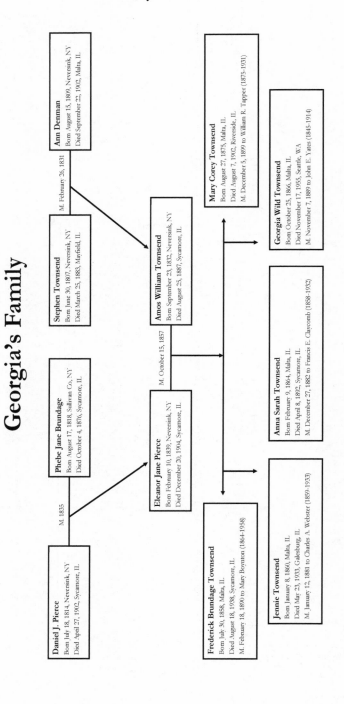

Jack's Family

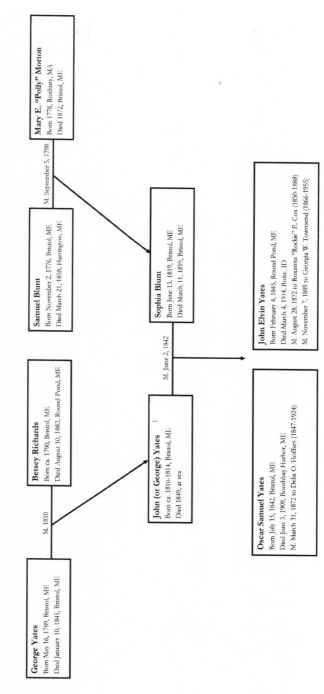

[1] See footnote 35 on page 36.

Georgia and Jack's Family

Georgia Wild Townsend
Born October 25, 1866, Malta, IL
Died November 17, 1955, Seattle, WA

M. November 7, 1889

John Elvin Yates
Born February 4, 1845, Round Pond, ME
Died March 4, 1914, Boise, ID

Dorothy Yates
Born January 30, 1891, Sycamore, IL
Died June 16, 1945, Seattle, WA
M. July 21, 1913 to Lee C. Morehead (1890-1915)
M. October 5, 1921 to Alford R. Halverson (1886- ??)

- **Penelope Pierce Morehead**
  Born June 5, 1914, Boise, ID
  Died September 30, 1951, Marysville, CA

- **Lee Clinton Morehead**
  Born October 30, 1915, Boise, ID
  Died February 25, 2009, Sacramento, CA

Margaret Yates
Born December 22, 1892, Chicago, IL
Died Unknown
M. August 25, 1915 to Frederick Breitinger (1888-1932)

- **Frederick William Breitinger, Jr.**
  Born May 8, 1916, Philadelphia, PA
  Died May 9, 1937, Philadelphia, PA

- **Margaret (Marge) Breitinger**
  Born April 19, 1919, Philadelphia, PA
  Died July 27, 1987, Camden, NJ

Marjorie Yates
Born December 22, 1892, Chicago, IL
Died January 21, 1963, Media, PA
M. December 14, 1917 to Isaac C. Cornog (1893-1971)

- **Douglas Yates Cornog**
  Born April 10, 1920, Philadelphia, PA
  Died January 5, 1985, Lee, FL

- **Geoffrey Yates Cornog**
  Born August 30, 1923, Philadelphia, PA
  Died July 21, 1997, Springfield, IL

John Townsend Yates
Born January 1, 1895, Round Pond, ME
Died March 8, 1895, Round Pond, ME

Oscar Townsend Yates
Born February 24, 1896, Round Pond, ME
Died October 18, 1957, Covington, KY
M. July 12, 1917 to Marian Pittinger (divorced in 1928)
M. 1930s to Helen S. Yates (maiden name unknown) (c.1897-1977)

Frederick Townsend Yates
Born May 13, 1901, Boise, ID
Died January 24, 1972, Seattle, WA
M. February 15, 1923 to Virginia C. Stewart (1905-1994)

- **Georgia Yates**
  Born April 19, 1925, Boise, ID
  Died August 25, 2011, Cashmere, WA
  M. March 18, 1946 to Paul D. Jones (1921-2004)

- **Frederick Townsend Yates, Jr.**
  Born April 7, 1931, Seattle, WA
  Died July 18, 2014, Brookings, OR

William Townsend Yates
Born October 11, 1904, Boise, ID
Died July 27, 1924, Boise, ID

Stephen Townsend Yates
Born November 26, 1907, Boise, ID
Died July 12, 1972, Seattle, WA
M. June 1, 1931 to Bernice H. Trotter (1908-1995)

- **Nancy Yates**
  Born October 18, 1937, Seattle, WA
  Died Unknown

Appendix B
Sea Shanties

Georgia would have heard the crew of the *Willie Reed* singing many sea shanties during the voyage. The two below are shanties she specifically mentioned in her letters. Sea shanties are traditional songs, so the lyrics are always different and evolving. Although the lyrics Georgia heard were almost certainly different than what is printed below, they would have been similar.

Whiskey Johnny

Oh, whiskey is the life of man
Whiskey, Johnny!
I drink whiskey when I can
Whiskey for my Johnny!

Whiskey from an old tin can
Whiskey, Johnny!
I'll drink whiskey when I can
Whiskey for my Johnny!

I drink it hot, I drink it cold
Whiskey, Johnny!
I drink it new, I drink it old
Whiskey for my Johnny!

Whiskey makes me feel so sad
Whiskey, Johnny!
Whiskey killed my poor old dad
Whiskey for my Johnny!

Whiskey killed my poor old dad
Whiskey, Johnny!
And whiskey drove my mother mad
Whiskey for my Johnny!

I thought I heard the old man say
Whiskey, Johnny!
I'll treat my crew in a decent way
Whiskey for my Johnny!

A glass of grog for every man
Whiskey, Johnny!
And a bottle full for the shanty-man
Whiskey for my Johnny!

Whiskey up and whiskey down
Whiskey, Johnny!
And whiskey all around the town
Whiskey for my Johnny!

Oh, whiskey here and whiskey there
Whiskey, Johnny!
Oh, I'll have whiskey everywhere
Whiskey for my Johnny!

Whiskey makes me pawn my clothes
Whiskey, Johnny!
And whiskey gave me this red nose
Whiskey for my Johnny!

Rio Grande [2]

I'll sing you a song of the fish of the sea
Way, Rio!
I'll sing you a song of the fish of the sea
And we're bound for the Rio Grande!

Then away, love, away
Way, Rio!
So fare ye well, my pretty young gal
We are bound for the Rio Grande!

It's goodbye to Sally and goodbye to Sue
Way, Rio!
And you who are listening, goodbye to you
For we're bound for the Rio Grande!

So man the good capstan and run it around
Way, Rio!
We'll heave up the anchor to this jolly sound
For we're bound for the Rio Grande!

---

[2] Also known as "Away, Rio!" and "Bound for the Rio Grande." In this shanty, "Rio Grande" refers to Rio Grande do Sul in Brazil, not the border river between Texas and Mexico.

> Then away, love, away
> Way, Rio!
> So fare ye well, my pretty young gal
> We are bound for the Rio Grande!

Our ship went a-sailing out over the bar
Way, Rio!
We pointed her nose for the southern star
For we're bound for the Rio Grande!

> Then away, love, away
> Way, Rio!
> So fare ye well, my pretty young gal
> We are bound for the Rio Grande!

The anchor is weighed and the sails they are set
Way, Rio!
The maids that we're leaving we'll never forget
For we're bound for the Rio Grande!

> Then away, love, away
> Way, Rio!
> So fare ye well, my pretty young gal
> We are bound for the Rio Grande!

Appendix C
A Tale of Two Transcripts

This project began with transcripts of Georgia's letters found in the collection of Doug Roberts, great-grandson of Georgia's brother, Fred Townsend. As mentioned before, Georgia asked her family to save the letters because they were the only account of her voyage. The family honored her wish, but no one living in Sycamore today knows what happened to the original letters, why the transcripts stayed in Sycamore after Georgia no longer lived there, or who transcribed the letters in the first place.[3]

A 1939 article from the *True Republican* mentioned that Georgia's niece Eleanor Townsend Roberts read some of the letters at a meeting of the Sycamore Unit of the Home Bureau. The article did not mention whether she read from the original letters or the transcripts. One theory floating around the family is that the original letters were thrown out after they were transcribed. At one time, this was an unfortunate but common practice, with descendants believing that typed versions of handwritten documents were the ultimate form of preservation. So in my mind, I gave up the original letters for lost, and I settled on working from the transcripts.

Three months into my research, however, I located a direct descendent of Georgia Townsend Yates, her great-granddaughter Wendy Jones Smoke of Wenatchee, Washington. To my delight, she had several family photos belonging to Jack and Georgia (many of them reprinted in this book). She also had several family heirlooms, including pieces from a tea set, Georgia's rocking chair, and Jack's sea chest that had traveled with him around the world. Most interesting of all, Wendy had her own set of transcribed letters.

I made the trip out to Washington to examine her collection, believing her transcriptions would be the same as mine (most likely a mimeograph or carbon copy). But as soon as I started thumbing through the brittle, onionskin pages, I realized her transcriptions were different. First, they were missing several entries on each end.[4] Second, it was clear to me that a different transcriber, someone who also had been working from the original letters, had typed them. So now I knew that the original letters

---
[3] Doug believes that his grandmother, Eleanor Townsend Roberts (Georgia's niece), had the original letters transcribed, though he is not certain.

[4] The set is missing the first thirteen entries (September 8, 1891 – October 18, 1891) and the last twenty-nine entries (August 2, 1892 – December 29, 1892).

had lived beyond Sycamore and made the journey west with Georgia. They had been passed down to her direct descendants, as she'd always intended.

Clues to the original letters' fate appeared in an undated, handwritten letter from Wendy's grandmother Virginia Stewart Yates (known affectionately as "Tutu"). Virginia was married to Frederick Yates, one of Jack and Georgia's five sons. She addressed the letter to her daughter, also named Georgia Townsend Yates, and explained that Georgia's sea letters had "seen the rounds through the family" and that several went missing after relatives plucked them from the stack as souvenirs. She also mentioned that she did not see the letters until after "Mothers Yates" had passed.

I got more of the story from Wendy's older sister, Virginia "Ginny" Snyder, who remembers her great-grandmother Georgia, having visited her several times before Georgia passed away when Ginny was seven years old. She gave me the following account as it had been passed down to her: Georgia had held on to the original letters herself, but at some point relatives asked whether they could publish them. Georgia was against this idea, because she didn't believe anyone outside the Yates family would have an interest in the letters. After Georgia passed, Virginia worried about what would happen to the rest of the letters. She feared relatives would continue picking them apart. At the time, she worked as a secretary at a liberal arts college, so she took the letters to work and transcribed them before any more were lost. She made one set, numbering the pages. Her set was also passed around to family members and many of those letters were lost as well, which is why the set that was passed down to Wendy Smoke begins on page six and is missing a total of forty-two entries.

So what happened to Georgia's original letters after Virginia transcribed them? Ginny speculated that Virginia tossed them out after she was finished, but I've learned from my earlier experience that this may not be the case. They could be out there still, tucked away in the collection of another family member.

Captain Jack Yates's sea chest, which accompanied him on his many sea adventures.

A ring and engraved napkin holders that belonged to Georgia.

This collection of portraits features Jack and Georgia and their seven children. It hung in Fred Yates's house and later in his daughter Georgia's house. According to Jack and Georgia's descendants, the collection was always referred to as the "Yard of Yates."

Appendix D
1912 Political Speech by Captain John E. Yates

During his 1912 campaign for state treasurer of Idaho, Captain John E. Yates gave the following speech on October 21 at the Pinney Theater in Boise. Captain Yates ran on the Progressive Party (or Bull Moose) ticket and used his speech to attack incumbent gubernatorial candidate James H. Hawley while promoting the Progressive gubernatorial candidate, G. H. Martin.

Reprinted from the October 22, 1912, issue of the *Boise Evening Capital News:*

> Ladies and Gentlemen: I know you have come here to hear our candidate for governor, Mr. Martin, and so I shall occupy but a few minutes, but I just want to say a few words to you.
>
> I was up here the other evening and heard Mr. Hawley speak from this stage, and he spoke, I thought, a little disparagingly of the legislature and its work, and as I was a member of the legislature, why I just want to say a few words about this matter.
>
> He spoke about his vetoes. He spoke somewhat boastingly about his vetoes. He said, I believe, he had vetoed about 30 bills and he was rather sorry he hadn't vetoed about 30 or 40 more. He spoke of his veto of the woman's property rights bill, and he had some kind of an excuse. One of the excuses I believe he had was that it interfered with the transfer of unpatented mining claims. I don't know whether ladies are much interested in mining or not, but that seemed to be about all of his excuse. He flew off on the divorce law, and that is about the end of it.
>
> Now, he vetoed another bill, and this was a bill to prevent young boys from frequenting pool rooms. Now, you fathers and mothers who have children—I have seven—you know the responsibility that is upon us to bring these young boys up, and it seems to me that if there is any law that could be passed or would be passed or was passed by the legislature that could keep these young boys out from the places whose ways take hold on Hell, it would be a pretty good law for us, for us fathers and mothers.
>
> (Applause.)
>
> Mind you, Mr. Hawley vetoed this bill.
>
> Then there was another bill that I remember. Now this was a bill designed for the protection of human life. I introduced that bill in

the senate. Now, the slaughter places in this country are the railroad crossings where these railroads cross at grade. Many hundreds of men are killed every year in this country upon railroad crossings, even those where the public country roads cross the railroads, but where steam and electric roads cross there is hardly a state in the union but has a law compelling them to come to a full stop. This law that I introduced there, which was passed by the senate and house, was designed for that purpose and compels steam and electric roads where they cross at grade to come to a full stop. Mr. Hawley vetoed this law. He said it would interfere with the business of railroads. Now, you husbands and you wives, your sons and your daughters, if they are brought home to you on a stretcher, dead or maimed for life, why you can get consolation from Mr. Hawley. He will tell you you haven't interfered with the business of railroads.

(Laughter and applause.)

Now, ladies and gentlemen, you don't have to take Mr. Hawley's pre-election promises, because you have got his record, and you don't have to take my pre-election promises because you have got my record. Mr. Hawley has been two years mayor of Boise; he has been two years governor of the state, and I have been in three sessions of the legislature and two years upon the Boise council. Now, you don't have to take our pre-election promises. You have got our records and if Mr. Hawley hasn't been faithful to his promises, if he hasn't done the very best he could for the people of this state, why don't you smoke him out.

(Laughter and applause.)

You have got a club, the ballot, and you hit him over the head with it.

Now, I have been in three sessions of the legislature. If I haven't fulfilled my duty there, if I haven't fulfilled the promises I made before I was elected the last time, that I would stand for the people and not for special interests, then you smoke me out, because I am a candidate for state treasurer on the Bull Moose ticket.

(Applause.)

Now, ladies and gentlemen, it has been said—it was in the morning paper, an editorial the other day—that we had no business upon the Progressive ticket. Now then, I want you to examine the record of all these men upon the Progressive ticket, and I will guarantee that you don't find a man there who has been wallowing in any sloughs, either social or political.

(Applause.)

Now, as far as business men are concerned, you go out upon the street, and if you find that I have been engaged in any crooked deal.

If you find that I have been engaged in organizing any joint stock companies and been a promoter and got a share of the promotion stock. If you find that I ever stood in with any contractors to defraud the state of Idaho—

(Applause.)

If you find I have been doing that, then you smoke me out. And if you find any of these other fellows on these other tickets have been doing these very same things, you smoke them out.

(Laughter and applause.)

Now, I don't want to occupy too much time, ladies and gentlemen, on this matter, but it seems to me I want to say a few words to you about the spirit prevailing in this campaign. Now, on one side there seems to be a good deal of vindictiveness and a good deal of, well, slander, and that sort of thing. Now, my friends, I think this was well illustrated in the two conventions held in the city of Chicago. At the national Republican convention, held in the city of Chicago, a great part of the resolutions and arts of that convention was received with the toot, toot of the steam roller, with groans of disapprobation and hisses. In the great Progressive convention held there, almost all of the resolutions were received with shouts of unanimous approval and with cheers and with cries of approval and songs of "Onward, Christian Soldiers."

(Applause.)

Now, I heard a man in the national Republican convention by the name of Bradley, Senator Bradley of the great state of Kentucky, attempt to make a speech. He hadn't proceeded very far when a man rose in the body of the convention and pointed his finger at him and said, "You voted for Lorimer." Instantly a murmur ran over the convention which increased until it became a roar of protest, and as far as having anything more, anything that that man said further, why his speech was done right there. At the national Progressive convention a man well known in this country, ex-Senator Beveridge, delivered the opening speech, defining the purposes of that convention and the Progressive party. When he closed his speech tears ran down the cheeks of bearded men, shouts and cheers rang through the convention hall; then someone began to sing, and this song was joined in until thousands were singing, till the great volume of song filled the vast auditorium, and some of the words they sang were these:

Mine eyes have seen the glory of the coming of the Lord,
He is trampling out the vintage where the grapes of wrath are stored.

Another part of it was:

> He has sounded forth the trumpet that shall never call retreat,
> He is sifting out the hearts of men before his judgment seat;
> Be swift, my soul, to answer him, be jubilant my feet,
> For God is marching on.

Now, my friends, these are inspiring words; aye, they are inspired words. This great hymn was written at the time when this country was in the throes of a great civil war and at a time when the soldiers of the north had met with reverses and the hearts of the friends of the union were heavy and depressed within them, but when this great song was published, a great wave of enthusiasm swept over the country, men everywhere sprang to arms for the defense of the union and the friends of liberty everywhere were filled with hope and courage. And so when this great hymn was sung in the national Progressive convention, it filled the hearts of those men and women assembled there with hope and courage. Many of the great newspapers of the country have characterized this as the camp meeting spirit. Now, I don't know about this camp meeting spirit, but it seems to me this is the spirit that stirs the hearts and souls of men and women; this is the spirit that stands for social and industrial justice; this is the spirit that stands for good government. So my friends, in this spirit, with strong hearts and undaunted courage, we face the dawn of a better day, which shall usher in a better government, a government for the people, by the people. So, my friends, in this spirit, we appeal to you to stand with the Progressive party for the principles it advocates, under the leadership of one of the greatest fighters, aye, one of the greatest champions of the rights of the people this country has ever seen, Theodore Roosevelt.

(Cheers and applause.)

Ladies and gentlemen, it will now be my great pleasure to introduce to you the Progressive candidate for governor, Mr. G. H. Martin.

(Applause and cheers.)

Appendix E

A Trip to Round Pond, Maine:
Excerpt from Amos Claycomb's 1902 Diary[5]

Amos Claycomb was born January 29, 1886, in Cameron, Illinois. He was the second of five children of Francis "Frank" Claycomb and Anna Townsend Claycomb (older sister of Georgia Townsend Yates). His family moved to the Daniel Pierce farm just outside Sycamore when he was two years old.

At the time of Amos's trip to Round Pond, Maine, he was sixteen years old. The route he took from Sycamore to Round Pond gives us an idea of the route Georgia, her mother, and their families would have taken when traveling between the two locales. His day-to-day activities provide a glimpse of what life was like in Captain Jack Yates's hometown, especially for all the Townsend family members who visited over the years.

Monday, July 7, 1902

It has been cloudy nearly all day and raining every once in a while.[6]

I started for Chicago at 11:07 this morning and at 8:15 P.M. I started for Maine.[7] I staid in Carson Pirie's most of the day.[8]

Tuesday, July 8, 1902

I rode on the cars all of last night and today until about two o'clock when we stopped until half past ten in Toronto, Canada. We rode around the city in the street cars and then took a ride on a boat over Lake Ontario over to an island for a little while.

Wednesday, July 9, 1902

We rode on the cars all night and arrived in Kingston, Canada early

---

[5] This journal excerpt is made possible by the dedicated work of Donna Catterick, a granddaughter of Amos Claycomb, who transcribed several years of Amos's diaries and made them available online at her genealogy website: This I Leave (http://thisileave.wordpress.com/).

[6] Amos began each entry with a description of the weather. After this first entry, I have removed these descriptions for brevity.

[7] Amos departed Sycamore on the Chicago Express train from the Chicago & North Western depot (corner of DeKalb Avenue and South Sacramento Street). The daily run left Sycamore at 11:07 a.m. and arrived in Chicago at 1:15 p.m.

[8] Carson Pirie Scott & Co. is a department store chain that began in Amboy, Illinois, in 1854 and had several stores in Chicago.

this morning. As the boat we intended to take did not arrive we had to stay here all day. At noon we all took a steamer and went down the St. Lawrence through the "Thousand Islands" which are very beautiful. We got home about six o'clock and ate supper at a hotel where we will stay all night.

Thursday, July 10, 1902

We all staid at a hotel all night and at six o'clock in the morning took a steamboat and went down the St. Lawrence River to Montreal where we took a train and will arrive at Portland early tomorrow morning. The boat we went down to Prescott on was called the Toronto and at that place we changed to another which was smaller and could shoot the rapids in the river. We ate breakfast on the Toronto a little lunch on the Bohemian and ate supper just before we got on the train in the depot.

Friday, July 11, 1902

Arrived at Portland in the morning and had to wait there until 12:40 P.M. when we took a train to Damariscotta and from there we came to Round Pond in a four-seated buggy getting here about supper time. I went out rowing for about ten or fifteen minutes after supper. At Portland I went up town for about an hour in the morning and bought a few things.

Saturday, July 12, 1902

I went hunting in the woods just back of the house this morning but did not see anything. Just after dinner I rowed the boat over to the town and got some lines and fish-hooks. I went fishing as soon as I come back for about fifteen minutes and got four fish, one a large-sized codfish. I then drove Grandma, Aunt Jennie and Aunt Mary out to the Cox farm and then to town.[9] After supper I rowed Louise and Marion over to town to get the mail.[10] Frederick, Charles and Pierce have been fishing and rowing all day.[11]

Sunday, July 13, 1902

I took a walk among the rocks this morning—when the water was at low tide—examining them. I went to church at two o'clock and staid to

---

[9] Amos's grandmother was Eleanor Pierce Townsend. His aunts were Jennie Townsend Webster and Mary Boynton Townsend, wife of Frederick Townsend. At this time, Amos's other Aunt Mary, Mary Townsend Tapper, was back in Illinois, suffering after a difficult childbirth the previous month. The Cox farm was most likely owned by a relation of Rockie E. Cox, the first wife of Amos's uncle Captain John Elvin Yates.
[10] These girls were Amos's sister Alta Louise Claycomb and his first cousin Marion Webster.
[11] These boys were Amos's first cousins Frederick C. Webster, Charles B. Townsend, and Pierce Webster.

Eleanor with sixteen of her grandchildren in 1903. Amos Claycomb stands in the upper right corner. (courtesy of Donna Catterick)

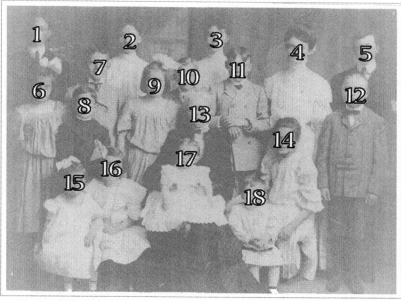

1. Frederick C. Webster 2. Alta Louise Claycomb 3. Eleanor Claycomb 4. Marion Webster 5. Amos Claycomb 6. Margaret Yates 7. Edward Claycomb 8. Oscar Yates 9. Marjorie Yates 10. Pierce Webster 11. George Claycomb 12. Charles Townsend 13. Eleanor Pierce Townsend 14. Dorothy Yates 15. Jane Tapper 16. Eleanor Townsend 17. Barbara Tapper 18. Frederick Yates

217

Sunday School eating dinner when I came back. I rowed over there with Charles and after dinner went out rowing with Pierce.

Monday, July 14, 1902

I got up a little early and rowed around the harbor for about an hour. After breakfast I picked about a quart of blueberries. I rowed over to town just before dinner to get the mail. After dinner I went fishing off Pulpit Rock and caught twenty-one rather small fish most of them cunners. Just before supper I rowed over to town to get a yeast cake. Fred, Charles and Pierce have been fishing almost all day. Aunt Mary, Aunt Jennie and Charles went out driving this afternoon.

Tuesday, July 15, 1902

I went rowing for three quarters of an hour before breakfast. Fished part of the time this forenoon and rowed Aunt Jennie over to town just before dinner to get some things. Caught two fish. Also went fishing this afternoon for a while and caught nothing. I started to get into the boat from the wharf tonight to row over after the mail and lost my balance and either had to go head or feet first. Chose latter and landed in three ft. water. Changed clothes and then went over with [no] other mishap.

Wednesday, July 16, 1902

I rowed a little this morning and over after the mail just before noon. I staid around the house most of the afternoon except to go with a man after some ice and also hoed in the garden for about an hour just before supper. Aunt Jennie, Aunt Mary, Louise, Marion and Charles drove over to town this afternoon. Fred and Pierce rowed over to town this forenoon and fished some of the afternoon. The ocean was a little rougher than usual.

Thursday, July 17, 1902

I hoed the garden nearly all the forenoon but went over after the mail at noon.

Grandma rented a cow yesterday and I began to milk her this morning. I fished and rowed this afternoon but only caught two fish. Marion, Louise and Fred went over after the mail tonight.

Friday, July 18, 1902

I went with Aunts Jennie and Mary, the girls and the boys to the porgie factory to fish but only caught four. I rowed over in the boat alone but Fred came back with me. I then fished a little near the old wharf and rowed some. Rowed over for mail just before dinner. Fred, Charles and I rowed over to town this afternoon and played ball for about two hours. I went fishing off Pulpit Rock for a while after supper and caught twenty fish making twenty five in all today. After this Fred and I went rowing for a few minutes.

Saturday, July 19, 1902

I went fishing this morning off Pulpit Rock with Fred and caught ten fish. He caught fifteen in the same time. We all went over to the rock for dinner and then I drove Grandma, Aunt Jennie and Aunt Mary around to make some calls. Grandma, Louise and Marion drove over to town this morning and got the mail. We received letters tonight tell[ing] us that Angeline Allen, my old nurse, died Tuesday night at eleven o'clock. The cause was the enlargement of the glands of the neck.[12]

Sunday, July 20, 1902

I walked a little over a half a mile this forenoon to get some chickens for dinner in the rain. I went both to church and Sunday School. Staid in the house rest of afternoon playing on Charles' gramophone. After dark tonight we boys took it down to the landing and played it for three sailors who came from a ship that was in the harbor to hear it. I rode in the buggy to church. I went out to a sailing boat to try to get some mackerel before breakfast but did not get any.

Monday, July 21, 1902

I put some Paris Green on the potato plants this morning but the rain afterwards washed it off.[13] I then went out rowing and also rowed after the mail just before dinner. I have staid in the house most of the afternoon but went after the mail just before supper.

Tuesday, July 22, 1902

I fished most of the morning and caught seventeen fish and again this afternoon catching nine making twenty six in all. I rowed over to the island

---

[12] In the back of the diary, Amos inserted an entry dated July 15, 1902: "At the home of William Young, Angeline died at the age of seventy-five years. She was my old nurse and has lived most of the time with Grandma Townsend for the last forty years. She died of a goiter which troubled her for a few months before her death."

[13] Paris Green was a powerful insecticide commonly used in this period.

this afternoon, which is just a mile from here and back again alone.[14] Louise and Marion rowed over after the mail both before dinner and supper. Fred fished for about an hour and a half this afternoon and caught fifty three fish.

Wednesday, July 23, 1902

I rowed first and then fished the rest of the forenoon catching fifteen fish. I fished most of the afternoon and caught fifty-five fish making seventy fish in all today. I learned to clean fish this afternoon.

Thursday, July 24, 1902

I went fishing this morning and caught twelve fish and cleaned some for dinner. I went over to town this afternoon and played ball most of the time.

Friday, July 25, 1902

I went fishing this morning and caught twelve fish. Aunt Jennie, Aunt Mary, Charles, Marion, Louise and I drove over to Damariscotta this afternoon and got back about eight o'clock tonight.

Saturday, July 26, 1902

I rowed over to town three times this morning and twice tonight. I went out driving with Grandma and Aunt Mary this afternoon. Aunt Jennie, Mrs. Sider, Marion, Louise and the three boys went on an excursion to Waldoboro this afternoon on the Steamer Medomak.

Sunday, July 27, 1902

I rowed over to town this morning before breakfast but have staid home the rest of the time. I went in bathing for a few minutes this forenoon but the water was too cold to have much fun. There was no church so we didn't go.

Monday, July 28, 1902

I have fished nearly all day but have only caught thirty fish. Grandma and some of the others drove over to town this afternoon and Louise and Marion went in bathing for a while. I went out on the schooner this morning which got caught on a rock as it was going out and staid on it talking with the sailors until it floated off when the tide became higher.

---

[14] He was referring to Louds Island, a four-mile-long, one-mile-wide island east of Round Pond. It was considered unincorporated territory after breaking with Bristol during a dispute over the 1860 presidential election.

Fred, Charles and Pierce also went in bathing today. Some men began to cut the hay around the house today.

Tuesday, July 29, 1902

I went in bathing this morning for about an hour and a half and then went over after the mail. I fished a little this afternoon and caught twelve fish. Mrs. Fuller and Mrs. Smith with her little boy came on the steamer at noon. I rowed over to the island tonight after supper and saw some fishermen setting out their nets for mackerel.

Wednesday, July 30, 1902

I got up early and took some mail over to town and fished a little before breakfast. I fished or rowed all the forenoon and caught twenty fish altogether. I dug a few clams and went in bathing for a little while this afternoon. About three o'clock we got a telegram that Aunt Mary Tapper was very sick. We packed up as quick as possible and drove to Damariscotta to catch the train that left here at 10:20 P.M.

Thursday, July 31, 1902

We arrived at Portland at one o'clock last night and staid in the depot until morning. We then took the train at 8:15 for Montreal and got there about seven o'clock. We then waited in the depot until about ten o'clock when we got on the train that will take us to Chicago. I went out walking around Montreal for a little while after we ate supper in the depot.

Friday, August 1, 1902

We started from Montreal last night and got into Chicago tonight about nine o'clock and were met by Papa, Uncle Will and Uncle Charlie.[15] All but Grandma, Aunt Jennie and Uncles Fred and Charlie came out to Sycamore.[16] We had to drive from DeKalb in buggies and we got home a little after two o'clock in the night. Grandma and the rest went out to Riverside to see Aunt Mary who is a very little better only there is not much hope yet.[17] We drove from DeKalb in two of our buggies.

---

[15] Papa was Amos's father, Francis "Frank" Claycomb. Uncle Will and Uncle Charlie were his uncles William Tapper and Charles Webster.
[16] Uncle Fred was Frederick B. Townsend.
[17] Amos's aunt Mary Tapper passed away the following Thursday, August 7, 1902.

Selected Bibliography

Archives and Collections:
Idaho State Historical Society. Boise, ID.
DeKalb County Archives. Joiner History Room. Sycamore, IL.
Mystic Seaport: Museum of America and the Sea. Mystic, CT.
Roberts Family Archives. Sycamore, IL.
Sycamore History Museum. Sycamore, IL.
Wendy Jones Smoke Collection. Wenatchee, WA.

Newspapers:
Alexandria (VA) Gazette
Baltimore American
Bangor (ME) Daily News
Bangor (ME) Daily Whig & Courier
Blackfoot (ID) Optimist
Boise (ID) Evening Capitol News
Boise, Idaho Daily Statesman
Boston Herald
Brooklyn Daily Eagle
Boston Traveler
Caldwell (ID) Tribune
Chester (PA) Times
Chicago Daily Inter Ocean
Chicago Daily Tribune
Daily Advertiser (Singapore)
Daily Kennebec (ME) Journal
Dallas Morning News
DeKalb County (IL) Republican Sentinel
DeKalb (IL) Daily Chronicle
Evening Post (Wellington)
Flakes Daily Galveston (TX) Bulletin
Indianapolis Journal
Indianola (TX) Weekly Bulletin
Lewiston (ME) Evening Journal
Los Angeles Herald
Maine Farmer
Monticello (NY) Republican Watchman

New York Evening World
New York Herald
New York Press
New York Sun
New York Tribune
Ogden City (UT) Evening Standard
Omaha Bee
Oregonian (Portland, OR)
Philadelphia Inquirer
Philadelphia Public Ledger
Portland (ME) Daily Press
Salem (MA) Register
Salt Lake Tribune
San Francisco Daily Alta California
San Francisco Morning Call
San Juan Islander (San Juan Island, WA)
Spirit Lake (IA) Beacon
Springfield (MA) Republican
Straights Times Weekly (Singapore)
Sycamore (IL) City Weekly
Sycamore (IL) Tribune
Sycamore (IL) True Republican
Tacoma (WA) Daily News
Velasco (TX) Times

Books and Articles:

American Lloyd's Register of American and Foreign Shipping. New York: American Lloyd's, 1873–1883.

Beasley, Nancy M. *The Underground Railroad in DeKalb County, Illinois*. Jefferson, NC: McFarland & Co., 2013.

The Biographical Record of DeKalb County, Illinois. Chicago: The S. J. Clarke Publishing Company, 1898.

Boies, Henry L. *History of DeKalb County, Illinois*. Chicago: O. P. Bassett, 1868.

Brassey, Annie. *The Last Voyage.* London: Longman's, Green, & Co., 1889.

Bristol, Maine Bicentennial, 1765–1965. Damariscotta, ME: News Print Shop, 1965.

Combination Atlas Map of DeKalb County, Illinois. Genoa, IL: Thompson & Everts, 1871.

Commercial Relations of the United States with Foreign Countries During the Years 1884 and 1885. Washington, DC: U.S. Government Printing Office, 1886.

Cordingly, David. *Women Sailors and Sailors' Women.* New York: Random House, 2001.

Creighton, Margaret S. and Lisa Norling, eds. *Iron Men, Wooden Women: Gender and Seafaring in the Atlantic World, 1700–1920.* Baltimore: John Hopkins University Press, 1996.

Dahlberg, Tonya M. "Our Family in DeKalb: Boynton, Pierce, Townsend." Unpublished manuscript, 2003. Roberts Family Archive, Sycamore, IL.

The DeKalb Chronicle Illustrated Souvenir. DeKalb, IL: J. F. Glidden Publishing Company, 1899.

French, Hiram T. *History of Idaho: A Narrative Account of Its Historical Progress, Its People and Its Principal Interests. Vol. 2.* New York: The Lewis Publishing Company, 1914.

The Genealogical Advertiser: A Quarterly Magazine of Family History, 1899. Vol. 2. Boston: Press of T. R. Marvin & Son, 1899.

Gordon, Thomas F. *Gazetteer of the State of New York: Comprehending Its Colonial History, General Geography, Geology, and Internal Improvements.* Philadelphia: T. K. and P. G. Collins, 1836.

Greene, Francis B. *History of Boothbay, Southport and Boothbay Harbor, Maine, 1623–1905.* Portland, ME: Loring, Short & Harmon, 1906.

Gross, Lewis M. *Past and Present of DeKalb County, Illinois.* 2 vols. Chicago: The Pioneer Publishing Company, 1907.

Hawley, James H. *History of Idaho: The Gem of the Mountains.* 4 vols. Chicago: The S. J. Clarke Publishing Company, 1920.

Hoyt, Elizabeth E. *Man and Nature in Bristol: A Contribution to the Bi-Centenary of the Town.* Damariscotta, ME: News Print Shop, 1965.

Johnston, John. *A History of the Towns of Bristol and Bremen in the State of Maine, Including the Pemaquid Settlement.* Albany, NY: Joel Munsell, 1873.

Manual, 1901: National Alliance of Unitarian and Other Liberal Christian Women. Boston: Press of George H. Ellis, 1901.

Matthews, Frederick C. *American Merchant Ships, 1850–1900.* Portland, ME: Southworth Press, 1930.

Murphy, Kevin C. *The American Merchant Experience in 19th Century Japan.* New York: RoutledgeCurzon, 2003.

Old Bristol Historical Society Blog; "Adventure and Romance on the High Seas to South Africa and Back," blog entry by Stanley Sawyer, 2014.

Outlook 55, no. 17 (Apr. 24, 1897): xi.

Portrait and Biographical Album of DeKalb County, Illinois. Chicago: Chapman Brothers, 1885.

Quinlan, James Eldrige, *History of Sullivan County: Embracing an Account of its Geology, Climate, Aborigines, Early Settlement, Organization; the Formation of Its Towns, with Biographical Sketches of Prominent Residents, etc., etc.* Liberty, NY: W. T. Morgans & Co., 1873.

Record of American Foreign Shipping. New York: American Shipmasters' Association, 1875–1894.

Rice, George Wharton. *The Shipping Days of Old Boothbay: From the Revolution to the World War with Mention of Adjacent Towns.* Portland, ME: Southworth-Anthoensen Press, 1938.

Round Pond as a Summer Resort. 1897.

Rowe, William Hutchinson. *The Maritime History of Maine: Three Centuries of Shipbuilding & Seafaring.* Freeport, ME: The Bond Wheelwright Co., 1948.

Sheehan, Beatrice Linskill. *Descendants of William Lain and Keziah Mather, with Her Lineage from Reverend Richard Mather.* Brooklyn, NY: Theo. Gaus' Sons, Inc., 1957.

Sketches of the Inter-Mountain States: Utah, Idaho, Nevada, 1847–1909. Salt Lake City: The Salt Lake Tribune, 1909.

Snyder, Ginny. Interview by author. Digital recording. Wenatchee, WA, Sept. 13, 2014.

Townsend, Eleanor P. "Diary." Unpublished manuscript, 1867. Wendy Jones Smoke Collection, Wenatchee, WA.

Unitarian Yearbook, 1910. Boston: American Unitarian Association, 1910.

United States Hydrographic Office, *East Indies Pilot, Volume 1: Island of Java, Islands East of Java, South and East Coasts of Borneo and Celebes Island.* Washington, DC: Government Printing Office, 1916.

Universalist Church Records, 1881–1903. Box 448, MS03. Joiner History Room, DeKalb County Archives, Sycamore, IL.

The Voters and Taxpayers of DeKalb County. Chicago: H. F. Kett & Co., 1876.

Witherell, Jim. *History Along the Greenbelt.* Caldwell, ID: Caxton Printers, 1992.

Acknowledgments:

Many people helped with the creation of this book: archivists, librarians, historians, family, and friends. I even picked up a few new friends along the way. First and foremost, this book would not exist without Doug Roberts, Georgia's great-grandnephew, whose family history intertwines with the history of Sycamore. Doug maintains an extensive family archive and brought Georgia's letters to my attention in the spring of 2014. His initial email served as the basic outline of this book. I must give a wholehearted thanks to Sue Breese, the director of the Joiner History Room, DeKalb County's historical archive. Sue answered hundreds of my emails and seems to know everything there is to know about Sycamore history. Any picture in this book credited to the Joiner History Room was located and given to me by Sue. She also maintains the Roberts Family Archive, which provided the bulk of the information in this book, including, most importantly, Georgia's letters. And also, thanks to Tom Matya, who cracked the whip when necessary, keeping me on task.

This book would exist in a much diminished form without the gracious help of Georgia's great-granddaughter Wendy Jones Smoke and her husband, Stan Smoke, who welcomed me into their home and treated me like part of the family. It was a pleasure to spend time with them, their children, and their grandchildren to see how Georgia and Jack Yates's unflagging curiosity, fortitude, and self-reliance has passed down through the generations. I am also grateful to the hospitality shown to me by Wendy's twin sister, Teresa "Toddi" Morneau, and her husband, Rich. And I must also thank Wendy and Toddi's older sister Ginny Snyder for her invaluable insight into Georgia's life and final years in Seattle. I hope I did your family history justice. If I left anything out, on a trotting horse it will never show.

I must also thank my editor, Kelly Unger, who makes me sound better than I am; Danielle Grundel, Archivist Technician at the Idaho State Historical Archives, who went above and beyond to track down some excellent images and information on the Yates family; and Herb Neu and Sasha Reynolds-Neu, who set me on the historical path. (Herb gets a double thanks, because he also handled the layout of this book.) I also received invaluable assistance from Maryjo McAndrew, Senior Archive Assistant at Knox College, and Pastor Dennis Johnson of the Federated Church of Sycamore.

And a quick shout out to all those who helped me both directly and indirectly (even if it was just to listen to me ramble on about something new I'd found): Rob Glover, director of the Joseph F. Glidden Homestead in DeKalb, Illinois; Michelle Donahoe, director of the Sycamore History Museum; Dan Templin, director of the DeKalb County Community Foundation; Jim Lyons; the Mesjaks (John, Laura, and Alice); Donna Catterick; Libby Harmon; Shane Sharp and Beth Schewe; Mark Pietrowski; Christine Brovelli O'Brien; and Aaron Fogleman, Anne Hanley, Beatrix Hoffman, Jim Schmidt, and the rest of Northern Illinois University's Department of History.

And as always, I thank my beautiful and talented wife, Gillian King-Cargile, who listened to me read all of Georgia's letters on a long drive to Alabama, and decided this was a project worth pursuing.

About the Author

Clint Cargile has worked as an English teacher, freelance writer, magazine editor, academic conference coordinator, landscaper, dish washer, car washer, dog washer, and veterinary assistant. He has a BA in English from the University of Alabama, an MFA in creative writing from Southern Illinois University, and an MA in history with a concentration in public history from Northern Illinois University. He is also the author of *Five-Mile Spur Line: A Railroad History of Sycamore, Illinois*. He lives with his wife and two daughters in DeKalb, Illinois. You can reach him at clintcargile@gmail.com.

Made in the USA
Middletown, DE
06 May 2017